Aziz Ibragimov
Abror Ruzmetov

AF482362

Past, current and future trends of SA/PHBA and their derivatives

(in chemistry branch)

2023 Year

Copyright © 2023 by A.Ibragimov, A.Ruzmetov.

All rights reserved. No part of this publication may be reproduced, distributed, or transmitted in any form or by any means, including photocopying, recording, or other electronic or mechanical methods, without the prior written permission of the publisher, except in the case of brief quotations embodied in critical reviews and certain other noncommercial uses permitted by copyright law. For permission requests, write to the publisher, addressed "Attention: Permissions Coordinator," at the address below.

ISBN 978-8-22393372-4 (Paperback)

First printing edition, 2023.

e-mail: uzchemist@gmail.com

TABLE OF CONTENTS

PART 1. BIBLIOMETRIC REVIEW TO THE SA/PHBA TRENDS FROM 2000 TO 2023

Theoretical and practical interest in salicylic and p-hydroxybenzoic acid and their derivatives has increased over the last two decades, and with it, academic study in the field has been burgeoning. Most scientometric studies have only focused to specific property of the topic compounds. None, however, are discussed in the origination progress and future prospects of SA and PHBA and their derivatives. The present study makes a bibliometric review of 2491 papers published during 2000-2023 which were indexed by the Scopus in the sub-discipline of salicylic/p-hydroxybenzoic and salicylates/p-hydroxybenzoates. Periodic trends in publishing, as well as prominent research sub-fields are detected, as are citation patterns, published journal areas of focus, major research institutions, funding organisations, significant research personalities, and the extent to which they interact with one another in research networks. Used keywords indicated the exact publication timeline. Before to 2010, the majority of research focused on the synthesis, structure, chemical and physical properties of salicylic/p-hydroxybenzoic acid and their derivatives. In the past decade, scientists have dedicated significant attention to identifying the bio-properties of the investigated substances.

1.1 INTRODUCTION

Hydroxybenzoic acids are considered polyphenolic acid [1], and these acids and their derivatives shows antipathogenic [2–5], antimicrobial [5–10], anticancer [11–14], antiviral [3, 15] and etc. It is known that everyday new compounds are synthesized by scientists and discovered their phy-chem-bio properties. But why this acid and its derivatives have not lost their actuality? Why are scholars in developed countries conducting experiments on these objects continuously?

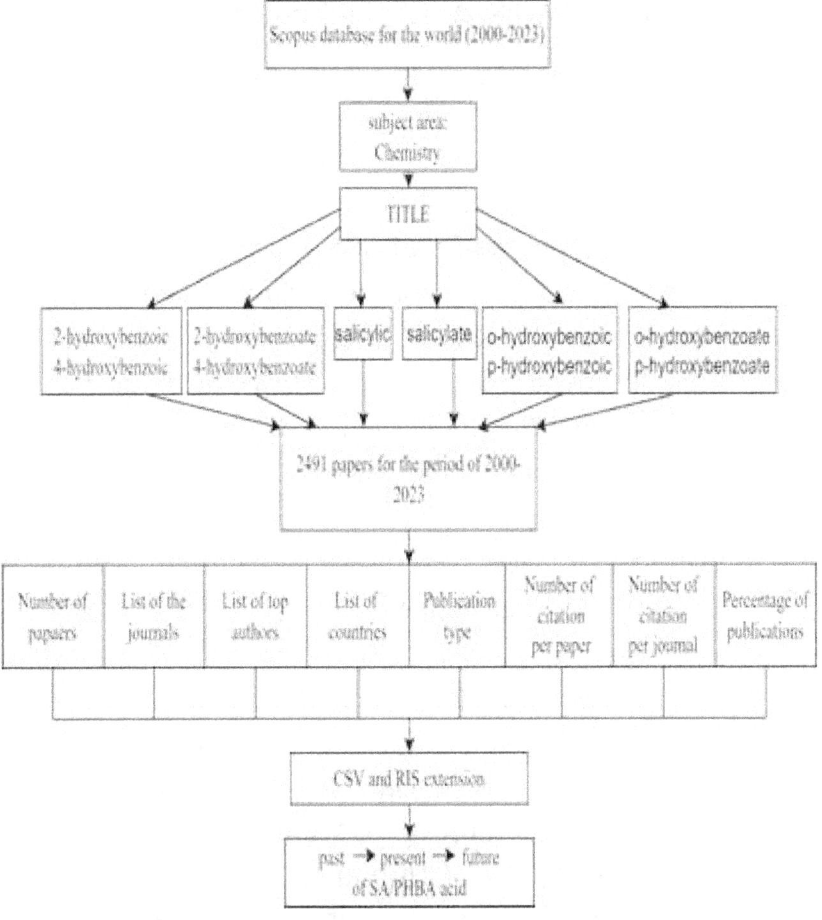

Figure 1. Flowchart of the methodology

Bibliometric is effective at providing researchers, stakeholders, and other policymakers with data that may be applied to enhance the quality of research [16]. Thus yet, no scientometric analysis of the trend of salicylic, p-hydroxybenzoic acid and their derivatives as chemical compounds has been performed. More thorough literature-based information will assist the community. Therefore, our research is intended to understand the past, present, and future of SA/PHBA acid and its derivatives as an important substance.

1.2 METHODS

Global library of SA/PHBA and their derivatives analyzed in the Scopus database for the period of 2000-2023 years. Firstly, we searched "salicylic" and 4-hydroxybenzoic as a keyword within article title, abstract and keyword section and found 34,765 and 2594 published documents respectively between abovementioned time frame. During skimming process, we identified that most papers are not fully devoted to chemistry of salicylic acid (SA)/p-hydroxybenzoic (PHBA) and their derivatives. Then applied following filters as a retrieval strategy on June 24, 2023: TITLE (salicylic OR salicylate OR 2-hydroxybenzoic OR 2-hydroxybenzoate OR o-hydroxybenzoic OR o-hydroxybenzoate OR 4-hydroxybenzoic OR 4-hydroxybenzoate OR p-hydroxybenzoic OR p-hydroxybenzoate) AND PUBYEAR>1999 AND PUBYEAR <2024 AND (LIMIT-TO (SUBJAREA, "CHEM")) AND (LIMIT-TO (LANGUAGE, "English")) AND (LIMIT-TO (PUBSTAGE, "final")) (**Fig 1**). By this refine we finally recorded 2491 papers and converted to CSV and RIS extension. VOSviewer (version 1.6.18; Centre for Science and Technology Studies, Leiden University: Leiden, The Netherlands) [17] is used to analyze bibliographic coupling, themes, co-authorship, co-occurrence, citation, and co-citations. English is a universal language, therefore, the literature in English is more standard and meaningful than literature in other languages.

1.3 RESULTS AND DISCUSSION

1.3.1. Published papers on SA/PHBA and their derivatives

In order to understand the development of *SA/PHBA and its derivatives* in chemical field, annual publication and citation trends were analyzed shown in **Fig. 2**. The publication on *SA/PHBA and their derivatives* research had come from 102 countries scattered all over the globe. The figure indicates that from 2000 to 2023, from 59 papers in 2000 to 152 papers in 2022, there is a general upward trend with a slight decrease in 2016-2020 years. We deduced that researchers are paying high concentrate on *salicylic and p-HBA acid and their derivatives* in chemical sector.

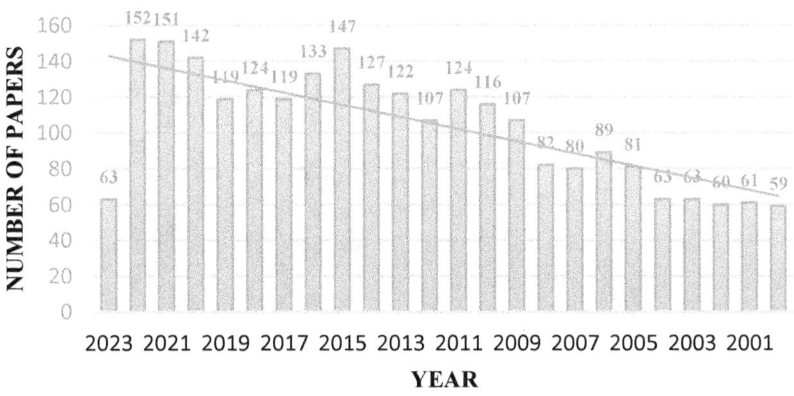

Figure 2. Number of papers on SA/PHBA and their derivatives by the year of publication in the world

1.3.2. Journals on SA/PHBA and their derivatives

A wide variety of journals in different parts of the world are used by scholars to publish their research. The communication patterns of the scholars indicate that the total output was distributed across 159 journals published in 145 countries. Of this 15 journals published 557 (21.5%) papers and remaining 78.5% papers were published in other journals. **Table 1** lists name of the 66 journals which published minimum 10 and higher number of papers during the abovementioned period. The second analyzing criteria is the name of publishing country and impact factor of the top 15 journals (**Table 2**). Of this five journals were published from the Netherlands, four of them from USA, three articles in UK and two papers from MDPI journals which publish in Switzerland. The average impact factor of the journals with the highest number of articles was 5.23. Among the 15 journals the Chemical Engineering Journal had the highest impact factor and the International Journal Of Molecular Sciences had the highest number of publications in this field.

Table 1. List of the top journals on SA/PHBA and their derivatives in the world

Scopus Source title	Number of papers	Scopus Source title	Number of papers
Int. Journal Of Molecular Sci.	60	Tetrahedron	14
Journal Of Molecular Structure	51	Tetrahedron Letters	14
Journal Of Molecular Liquids	43	Bulletin Of The Kor. Chem Society	13
Acta Crystallographica Section E	42	Chemical And Pharm. Bulletin	13

Bioorganic And Med. Chemistry	15	Carbohydrate Polymers	10
Biosci. Biotech. And Biochem.	15	Journal Of Chem. Thermodynamics	10
Electrochimica Acta	15	Journal Of Crystal Growth	10
Jour. Of Ther. Analy. And Calor.	15	Journal Of Liquid Chromatography	10
Russian Journal Of Gen. Chem.	15	Organic And Biom. Chemistry	10
Langmuir	14	Rasayan Journal Of Chemistry	10
Physical Chemistry Chem. Phys.	14	Synthetic Communications	10

1.3.3 Top authors on SA/PHBA and their derivatives

There is no doubt that authors play an important role in the development of a particular field of research in any region of the world. By analyzing authors with the most publications, this section provides insight into the research contributions of individuals. **Table 3** illustrates total of 15 different top authors worked to publish papers for the chemistry of SA/PHBA and their derivatives during the last two decades. According to statistical information, Khan is a top contributing author with 27 publications from University of Malaya (Malaysia), whereas Huang J. and Rasmuson A.C. are the most cited researchers. Regardless of 9 papers, Wang, X. is cited 31.3 times for each his works. Huang J., Khan M.N. and Rasmuson A.C. have the most publications and higher H-index, which extends their position as the leading experts in SA/PHBA and their derivatives in the chemical field.

Table 2. Distribution of research output in prolific journals

Journal	TNP (%)	Publishing country	IF
International Journal Of Molecular Sciences	2.18%	Switzerland	6.01
Journal Of Agricultural And Food Chemistry	1.79%	USA	5.5
Rsc Advances	1.74%	UK	3.73
Journal Of Molecular Liquids	1.69%	Netherlands	6.21
Spectrochimica Acta Part A Mol. And Biom. Spectroscopy	1.39%	Netherlands	4.83
Chemical Engineering Journal	1.34%	Netherlands	16.74
Food Chemistry	1.34%	UK	9.23
Journal Of Molecular Structure	1.34%	Netherlands	3.84
Asian Journal Of Chemistry	1.24%	India	0.49
Journal Of Physical Chemistry A	1.24%	USA	2.72
Crystal Growth And Design	1.14%	USA	3.69
Journal Of Chemical And Engineering Data	1.14%	USA	3.1
Molecules	1.14%	Switzerland	4.927
Acta Crystallographica Section E Structure Reports Online	1.04%	UK	0.91
Talanta	1%	Netherlands	6.55
Total:	**21.5%**		

Table 3. List of top productive authors in research on SA/PHBA and their derivatives issue in the world

No.	Author	TP(R)	H-index (by topic theme)	Total citations	Avg citation/item	T.C./ year
1.	Khan, M.N.	27	12	278	23.2	12.64
2.	Dega-Szafran Z.	17	9	202	11.9	8.42
3.	Langer, P.	17	6	147	24.5	9.8
4.	Huang, J.	14	14	385	27.5	35
5.	Ismail, E.	14	10	209	20.9	9.5
6.	Rasmuson, A.C.	12	12	318	26.5	19.9
7.	Yan, Y.	11	10	223	22.3	18.6

8.	Fischer, C.	9	8	62	7.75	4.13
9.	Meng, M.	9	8	139	17.4	15.44
10.	Wang, X.	9	7	219	31.3	19.9
11.	Gurina, D.L.	8	5	47	9.4	3.36
12.	Yuan, R.	8	5	69	13.8	3.63
13.	Antipova, M.L.	7	5	44	8.8	6.3
14.	Furia, E.	7	5	87	17.4	4.58
15.	Lubes, V.	7	4	30	7.5	2.5

1.3.4 Top countries on SA/PHBA and their derivatives

A region's internationalization can be measured by the number of countries participating in research activities on a particular topic. Total of more than 96 countries jointly worked on SA/PHBA and their derivatives in the world from 2000-2023. **Figure 3** highlights geographical location of the top 15 countries participated in publishing of at least 45 and more publications. On the figure with huge difference China holds a dominant by 600 research documents representing 23.3% of overall publications, followed by India (345; 13.3%), the USA (222; 8.6%), Iran and Japan are almost same level (~ 5.3%). Researchers from these top five countries participated in publication of 59.4% of total papers.

1.3.5 Top institutions on SA/PHBA and their derivatives

Generally, institutions are ranked as a result of the quality of the papers published by their researchers. 160 different institutions worked in cooperation to publish 2010 papers on SA/PHBA and their derivatives over the world the period of 2000-2023. In order to identify the most influential and productive institutions in SA/PHBA and their derivatives chemistry, we have analyzed the top 15 institutes' publications.

As indicated in **Fig. 4**, of the 15 institutions, six of them were from China, two of them from Germany, one each from Malaysia, Russia, France, Germany, Poland, Brazil and Iran. These prolific institutions contributed around 16.9 % of the total output. Among these, the Ministry of Education China occupies the 1st position in record rank (67 records), followed by Russian and Chinese Academy of Sciences 58 and 56 records respectively.

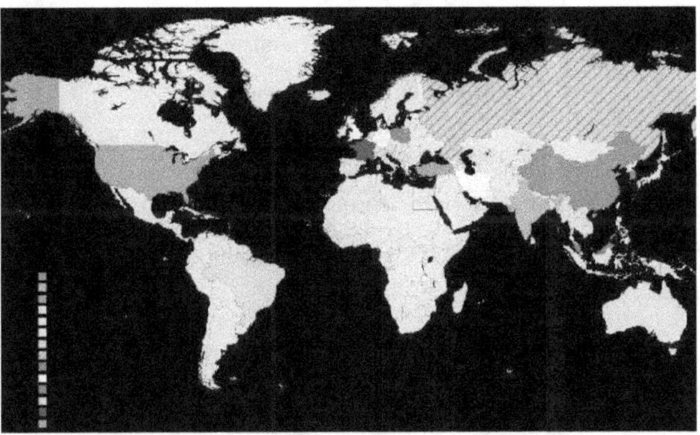

Figure 3. List of top 15 countries on SA/PHBA and their derivatives in the world

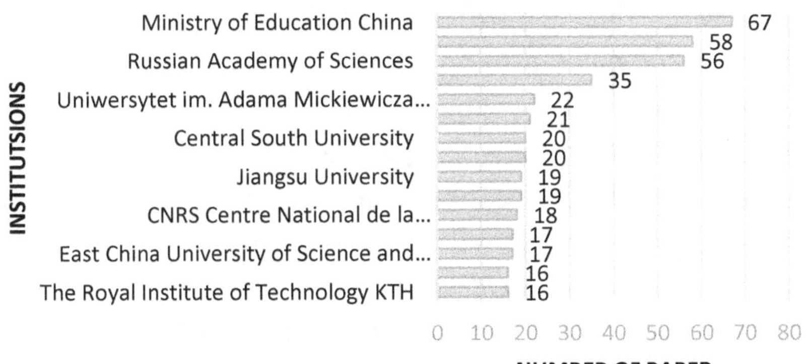

Figure 4. List of top institutions on SA/PHBA and their derivatives issue in Central Asia

1.3.6 Most Important Funding Sponsors

A funding sponsor is an organization or program that provides funding to researchers for the purpose of conducting research. Several of them are related to the chemistry of SA/PHBA and their derivatives (**Figure 5**).

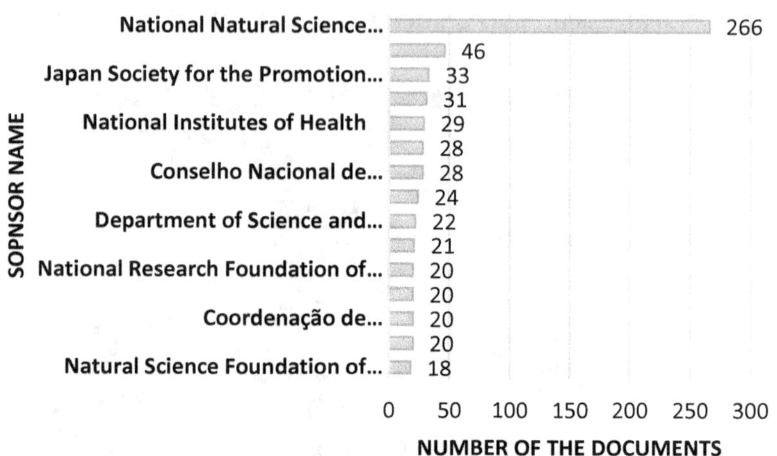

Figure 5. Main funding sponsors in SA/PHBA and their derivatives

Thus, we will start scrutinizing these systems according to the number of published articles. National Natural Science of Foundation of China is in the first position with 266 articles, and second place would go to the National Science Foundation, but at a great distance from the above, with 46 articles published about the chemistry of SA/PHBA and their derivatives. The Japan Society for the Promotion of Science is close behind, with 33 articles. Authors [18] remarks such activity of the NSFC because of the funding improvement - longer duration and larger grant - has resulted in an increase in research output, as measured by the number of publications.

By country, there are four funding sponsors from China, representing one third of this ranking. It is followed by the USA with three institutions or programmes. The European Union (EU), Brazil, Japan and India have two funding sponsors.

1.3.7 Top cited papers on SA/PHBA and their derivatives

There is a correlation between the number of citations and the quality and novelty of the research conducted. Fifteen mostly cited papers on SA/PHBA and their derivatives shown in the **Table 4** [19–34]. Total of 45,543 citations given to 2491 publications on SA/PHBA and their derivatives for the given period. These 15 documents are cited 2897 times and represents 6.36% of total citations. Of these 15 papers, and China have 5, USA have four representations, Spain reserves second place with three corresponds whereas Iran, Ireland, Russia and Poland have one participant each in the list of most cited work. Surprisingly, in spite of 13th position, L.J. Huang`s paper named after "Signaling crosstalk between salicylic acid and ethylene/Jasmonate in plant defense: Do we understand what they are whispering?" had highest annual citation trend with 51 citations per year. There is almost no difference between IF of the journals of Organic Letters (6.06) and International Journal of Molecular Structures (6.01, MDPI). There are several possible explanations for this result. Firstly, the Organic Letters is not free access journal [35] and in this case MDPI journals might have attracted more attention of scientists with its full accession. Secondly, R. Shen`s paper is devoted to synthesizing sections only. However, L.J. Huang and others discussed both experimental and bio-properties on the topic and this factor may have made the article more readable and informative.

Table 4. List of top cited publications on SA/PHBA and their derivatives in the world

№	Title	Journal	Corresponding author	Cntry	PY	TC	TC/Y
1.	Sorptive removal of salicylic acid and ibuprofen from aqueous solutions using pine wood fast pyrolysis biochar	Chem. Eng. Journal	T.E.Mlsna	USA	2015	264	29.3
2.	Unconventionally secreted effectors of two filamentous pathogens target plant salicylate biosynthesis	Nature Comm.	D. Dou	China	2014	239	23.9
3.	Synthesis of enamides related to the salicylate antitumor macrolides using copper-mediated vinylic substitution	Org. Letters	R. Shen	USA	2000	236	10.7
4.	Lanthanide oxide doped titanium dioxide photocatalysts: Effective photocatalysts for the enhanced degradation of salicylic acid and t-cinnamic acid	Journal of Catalysis	K.T.Ranjit	USA	2001	229	10.9
5.	Molecular ladders with multiple interpenetration of the lateral arms into the squares of adjacent ladders observed for [M2(4,4'-bpy)3(H2O)2(phba)2(NO3)2·4H2O (M = Cu2+ or Co2+; 4,4'-bpy = 4,4'-bipyridine; phba = 4-hydroxybenzoate	Inorganic Chemistry	Tong M.	China	2000	197	8.56

No.	Title	Journal	Author	Country	Year		
6.	Vapour treatments with methyl salicylate or methyl jasmonate alleviated chilling injury and enhanced antioxidant potential during postharvest storage of pomegranates	Food Chem	D. Valero	Spain	2011	176	16
7.	Effect of pre- and postharvest salicylic acid treatment on ethylene production, fungal decay and overall quality of Selva strawberry fruit	Food Chem.	M. Asghari	Iran	2007	173	11.53
8.	In search of pure liquid salt forms of aspirin: Ionic liquid approaches with acetylsalicylic acid and salicylic acid	Phy.Chem. Chem.Physics	R. D. Rogers	Ireland	2010	166	15.1
9.	Photocatalytic deactivation of commercial TiO2 samples during simultaneous photoreduction of Cr(VI) and photooxidation of salicylic acid	Jour. of Photochem.and Photobio. A: Chemistry	J.A.Navio	Spain	2001	161	7.67
10	The synthesis of chromenes, chromanes, coumarins and related heterocycles via tandem reactions of salicylic aldehydes or salicylic imines with α,β-unsaturated compounds	Org. and Biomolecular Chem.	M. Shi	China	2007	158	10.53
11	Proteome approach to characterize proteins induced by antagonist yeast and salicylic acid in peach fruit	Journal of Proteome Research	S. Tian	China	2007	157	10.47

12	Salicylic acid receptors activate jasmonic acid signalling through a non-canonical pathway to promote effector-triggered immunity	Nature Comm.	X. Dong	USA	2016	156	26
13	Tuning gold nanorod synthesis through prereduction with salicylic acid	Chemistry of Materials	L.M. Liz-Marzán	Spain	2013	156	17.33
14	Signaling crosstalk between salicylic acid and ethylene/Jasmonate in plant defense: Do we understand what they are whispering?	Int. Journal of Molecular Sciences	L.J. Huang	China	2019	153	51
15	Task-specific ionic liquid trioctylmethylammonium salicylate as extraction solvent for transition metal ions	Talanta	I.V. Pletnev	Russia	2010	153	13.91

1.3.8 Top cited journals on SA/PHBA and their derivatives

A minority of scientific journals publish the majority of scientific papers and receives the majority of citations [36]. In 3.2 subsection we analyzed top journals which published most number of papers. However, despite of fever number of papers there are journals that gathered more citation from 2000 to 2023. Taking into consideration this factor we decided to investigate top-cited journals on SA/PHBA and their derivatives in the chemical branch. Firstly, we sorted source names alphabetically of excel extension file of 2010 documents. Then step by step total papers` citations are summarized by each journal. Interestingly, at the result we got updating list with potential journal names. Initial 15 journals selected and shown in **Figure 6**.

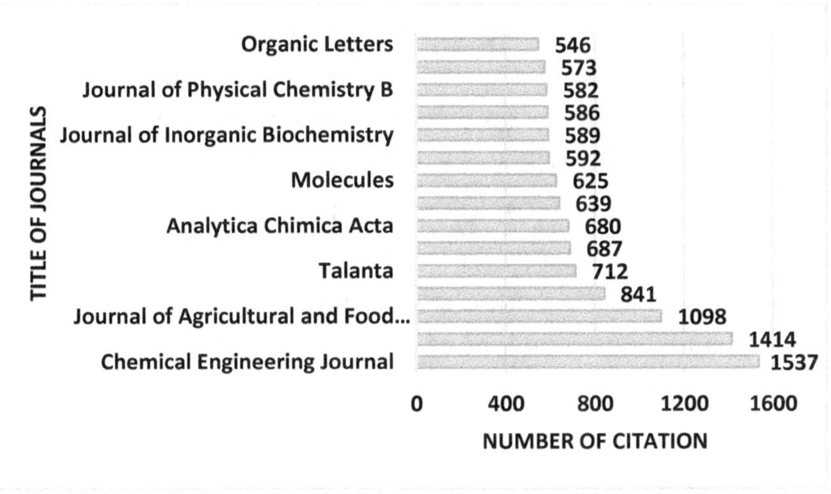

Figure 6 List of top cited journals on SA/PHBA and their derivatives in the world.

Almost 32% (11701 citations) of total citations given to papers published in these 15 journals. As a result of the number of citations, the 6th ranked journal with 27 documents by publication rate emerged as the best journal (**Table 1, 2**). Instead of five 1-5 journals in the **Table 2**, Analytica Chimica Acta, Bioorganic and Medicinal Chemistry, Journal of Physical Chemistry B, Organic Letters, Journal of Inorganic Biochemistry are ranked as a top cited journal on chemistry of SA/PHBA and their derivatives.

1.3.9 Publications by keywords on SA/PHBA and their derivatives

Co-authorship, keyword co-occurrences, citations, bibliographic coupling, and co-citation maps can be generated using VOSviewer based on bibliographic data. File formats supported include .txt, .ris, and .csv from databases such as

Web of Science, Scopus and PubMed [37]. The raw file was imported into VOSviewer and a map of keyword co-occurrences (**shown in Figure 7**) was created using the software. There are 1000 items distributed over five clusters: cluster 1 (407 items), cluster 2 (217 items), cluster 3 (167 items), cluster 4 (164 items), cluster 5 (45 items). When we divide the period into 3 time lapses, used keywords showed actual chronology of publishing directions.

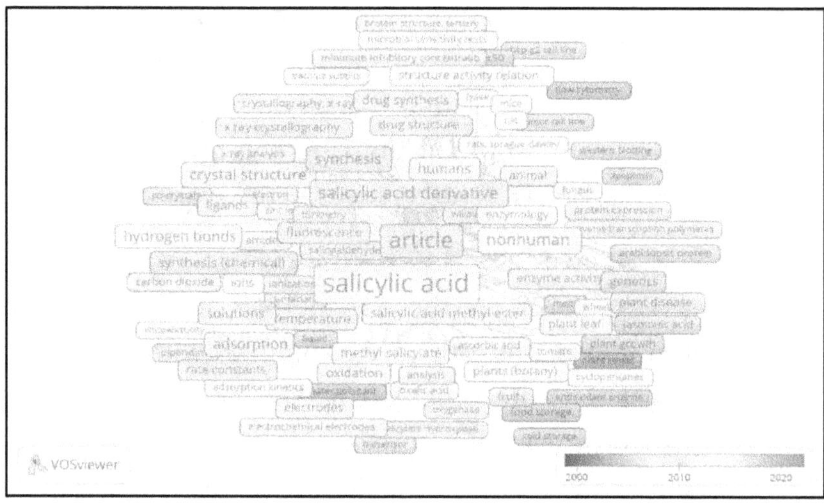

Figure 7. Keyword co-occurrences VOSviewer analysis based on Scopus

The synthesis, structure, chemical and physical properties of SA/PHBA and their derivatives are studied mainly until 2010. In the last decade, much attention has been paid by scientists to the identification of the bio-properties of the studied compounds. This may have occurred because wide investigating has been done on structural, chemical and physical parameters of the topic compounds.

1.3.10 Visualization of co-authorship countries

International collaboration is becoming increasingly important in scientific research. It is **Figure 8** demonstrates a VOSviewer (version 1.6.18) of visualization of the 60 most co-authorship countries in eight clusters. The minimum number of documents of a country and number of citations of a country is set as 3. As demonstrated in **Fig. 7**, the co-authorship study of countries illustrates the collaboration relationship among countries throughout this subject, as well as the level of cooperation. The thicker and longer links between nodes show cooperative interaction between countries, while the larger nodes represent the most productive nations.

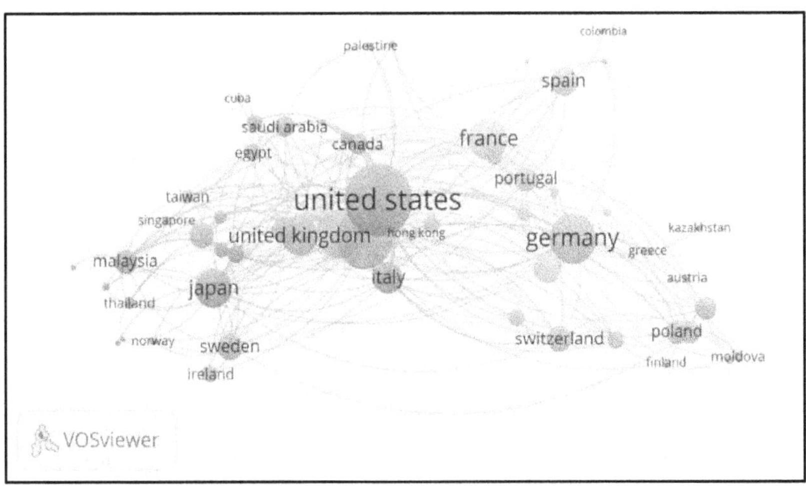

Figure 8. Structure map of countries co-authorships on SA/PHBA and their derivatives

The countries with the highest total link strength (TLS) were the USA with 178 documents and the link strength of 601 (collaborated with 35 countries) followed by India with 78 TLS (collaborated with 28 countries) and 281 documents. Germany was in the third position with 98 documents and TLS of 62 (collaborated with 28 countries). Regardless of 448 papers, China occupied the 4th position with a link strength of 60 and 25 collaborating ties from 2000 to 2021 (**Table 5**).

Table 5. Countries co-authorships on SA/PHBA and their derivatives

Country	Docs	TLS	Country	Docs	TLS	Country	Docs	TLS
United States	182	102	S. Arabia	20	19	Greece	11	6
India	282	78	Belgium	15	15	Romania	14	6
Germany	98	62	Pakistan	24	15	Slovenia	9	6
China	451	60	Egypt	37	14	Austria	7	5
France	55	49	Iran	107	14	Indonesia	12	5
Japan	108	44	Ireland	16	14	Norway	7	5
United Kingdom	67	44	Netherlands	16	14	Palestine	6	5
Italy	53	33	South Africa	16	14	Argentina	17	4
Czech Republic	20	30	Mexico	20	13	Bangladesh	3	4
Spain	75	29	Slovakia	12	13	Israel	6	4
Australia	29	27	Turkey	39	12	Tunisia	5	4
Sweden	23	26	Taiwan	39	11	Vietnam	4	4
South Korea	51	25	Brazil	45	10	Cuba	3	2
Switzerland	11	25	New Zealand	7	10	Hong Kong	5	2
Malaysia	60	23	Thailand	24	9	Hungary	10	2
Poland	64	23	Moldova	7	8	Kazakhstan	4	2
Portugal	27	23	Serbia	12	8	Nigeria	6	2
Russian Federation	80	21	Croatia	7	7	Algeria	8	1
Ukraine	27	21	Singapore	7	7	Colombia	4	1
Canada	28	19	Finland	6	6	Venezuela	8	0

1.3.11 Co-Authorship analysis of authors

There were 6,591 authors who participated in the publication of SA/PHBA and their derivatives. In this survey, all authors filtered at least 3 announced papers and a total 468 authors were found automatically by VOSviewer. Instead of all items the 235 largest set of connected items were showed to measure total link strength. The 14 most significant clusters of total authors were categorized. Largest number of cluster consisted of 30 authors and the last one ends with 8 authors. Calculated links equal to 1485, this means that the articles published jointly by 235 scientists (at least two authors) amounted to 1485. The **Figure 9** demonstrates density visualization of total link strength of co-authorship collaborations.

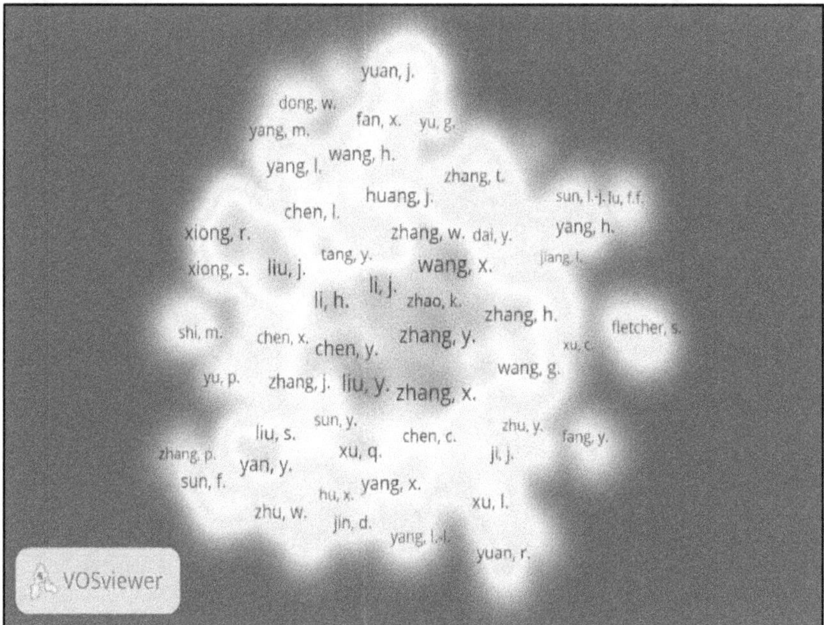

Figure 9. Density visualization map of co-authorship

25

1.4 CONCLUSION

The chemistry of SA/PHBA and their derivatives are crucial topic for science and engineering sector. Based on the Scopus database we tried to analyse papers published on abovementioned compounds in English for the whole world during 2 decades by bibliometric method. Developing countries, especially China, have more publications, institutions and funding sponsors. Several industrial nations, such as the United States, India, Japan, and Malaysia, have progressed well, however there is still a significant gap between China and the rest of the countries. Based on the current trend in citations to published manuscripts, we conclude that researchers will continue to investigate the bio-features of the title compounds for usage in the agricultural, food, and manufacturing industries.

PART 2. PARTICULAR PROPERTIES OF MONONUCLEAR METAL COMPLEX COMPOUNDS OF SA/PHBA

2.1 CRYSTAL STRUCTURE OF THE BIS(SALICYLATO)-DIAQUA-ZINC (II)

2.1.1 Introduction

Salicylic acid belongs to the carboxylate ligands, and depending on the chemical properties of the metal atom also the reaction conditions, it realizes the different modes of the coordination [38]. In many cases it forms the chelate cycle coordinating the metal ion through O atoms O of the COO- and OH-groups. The coordination compounds of d-metals with the different isomers of hydroxybenzoic acid are widely stud- ied by the single crystal X-ray methods [5].

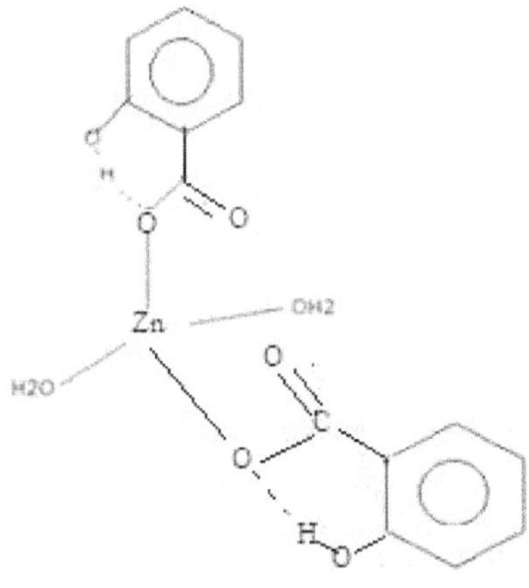

Fig 1. Chemical structure of $[Zn(Sal)_2(H_2O)_2]$

We plan to prepare an ointment on the bases of the coordination compound of the zinc with salicylic acid and because this complexes have been synthesized and its crystal structure was studied. Primary experi- ments show that the ointment composed from coordination compound zinc - salicylate ligand is more efficiently than the ointment composed from zinc and it may be used in the preparation of the skin care ointment.

2.1.2 Methods and materials

All reagents were readily available from commercial sources and were used as received without futher purification. Analysis of C, H and O were performed on a German ElementarVario EL instrument. Data for the crystal structure determinations were collected on an Oxford Diffraction Xcalibur-R CCD diffractometer (*Cu*Ka-radiation, λ=1.54184 Å, ω-scan mode, graphite monochromator, at 293K)[39]. The structure were solved by a direct method of the SHELX-97 program package [40, 41]. All non- hydrogen atoms were refined anisotropically. Hydrogen atoms were inserted at calculated positions and constrained with isotropic thermal parameters. The molecular drawings were plotted by MERCURY program package[42]. For The crystallographic data and details of the structure refinement are given in**Table 1** and the coordinates of the atoms on the **Table 2**.

Mononuclear complex bis(salycilato)-diaqua-zinc, [Zn(Sal)$_2$(H$_2$O)$_2$] [I] (where Sal is salycilato anion with deprotonated carboxylate group) has been synthesized in the water solution by mixing of the zinc acetate and salicylic acid in molar relation Zn : HSal=1:2. The single crystals of the complex I have been prepared by slowly evaporation of the mother solution. Analysis of C, H and O were performed on a German Elementar Vario EL instrument.

Table 1. Crystal and experimental data

Empirical formula	$C_{11}H_{10}NOZn$
1	2
Formula weight, g/mole	250.44
Temperature, K	298
Crystal system	monoclinic
Space group	C2
a, Å	15.490(1)
b, Å	5.348(1)
c, Å	9.179(1)
a,°	90
β,°	93.53(1)
y,°	90
Volume, Å³	758.9(1)
Z	3
ρcalcg/cm³	1.64
μ/mm⁻¹	2.633
F(000)	382.2
Crystal size/mm³	0.2 × 0.2 × 0.3
Radiation	Cu Ka (λ = 1.54184)
2Θ range for data collection,°	11.44 to 150.52
Index ranges	$-15 \leq h \leq 19$, $-5 \leq k \leq 5$, $-10 \leq l \leq 11$
Reflections collected	1088
Independent reflections	860 [R_{int} = 0.0135, R_{sigma} = 0.0336]
Data/restraints/parameters	860/1/108
Goodness-of-fit on F^2	1.061
Final R indexes [I>=2σ (I)]	R_1 = 0.0405, wR_2 = 0.1049
Final R indexes [all data]	R_1 = 0.0414, wR_2 = 0.1082
Largest diff. peak/hole / e Å⁻³	0.26/-0.25
Flack parameter	0.09(6)

Table 2. Fractional Atomic Coordinates ($\times 10^4$) and Equivalent Isotropic Displacement Parameters($Å^2 \times 10^3$) . Ueq is defined as 1/3 of of the trace of the orthogonalised Uij tensor.

Atom	x	y	z	U(eq)
Zn1	5000	1890.4(11)	5000	46.2(4)
O3	7516(2)	5361(8)	3906(4)	54.9(11)
O2	5999(2)	3502(9)	4095(4)	52.2(10)
O1W	4244(3)	-526(9)	3860(6)	69.4(13)
O1	4862(2)	5183(9)	3008(5)	59.8(11)
C1	6242(2)	6921(18)	2530(4)	41.4(10)
C2	7136(2)	7013(19)	2923(4)	43.7(11)
C7	5660(3)	5113(12)	3218(5)	45.1(14)
C3	7657(3)	8759(13)	2306(6)	52.5(15)
C6	5894(4)	8624(14)	1505(5)	54.4(15)
C4	7293(4)	10429(14)	1299(6)	57.9(16)
C5	6414(4)	10381(15)	902(6)	64.9(19)

2.1.3 Results and discussions

The crystals is monoclinic, space group C2, with a=15.490(1), b=5.348(1), c=9.179(1) Å, β=93.53(1)°, V=758.9(1) Å3, dcalc= 1.64 g/cm3, Z=3. The factors R1[I≥2σ(I)] and wR2(all data) values are 0.0405 and 0.0414, respectively, for all 860 independent reflections. The crystal structure is built from mon- onuclear molecules of Zn [(Sal₂)(H₂O)₂. The atom Zn is located at the symmetry axis 2 and has tetrahedral coordination polyhedra involving two O atoms of the two symmetrically relaited monodentate Sal-ligands and two water molecules. The carboxylate group coordination mode is the syn-1a monodentate type.

In crystal I, *Zn* atom is located at the crystallographic axis 2 and surrounded on the tetrahedral mode by the O atoms of carboyalte group of two monodentate *sal*-ligands and two water molecules (**Fig.1**). The carboxylate group coordination mode is the syn monodentate (1a). The angle C7-O2-*Zn* equal to 104.8(2)°.

Table 3. Selected bond distances (d, Å) and angles (°, grad.)

Bond	d	Bond	d
1	2	3	4
Zn1-O2	1.997(3)	Zn1-O2	1.997(3)
Zn1-O1W	1.996(5)	Zn1- O1W	1.996(5)
O3-C2	1.370(8)	O2-C7	1.271(7)
O1- C7	1.240(7)	C1-C2	1.409(5)
C1- C7	1.489(9)	C1- C6	1.394(9)
C2- C3	1.379(10)	C3- C4	1.381(10)
C6- C5	1.377(10)	C4 -C5	1.388(10)
O2 Zn1	128.9(3)	O1W Zn1	120.41(19)
O2 O1W	120.41(19)	O2 O1W	93.02(18)
Zn1 O2	93.02(18)	Zn1 O2	99.3(3)
O1W Zn1	104.8(3)	O1W Zn1	121.7(7)
O2C7 O2	117.9(4)	O1WC7	120.7(7)
Zn1 C3	119.2(5)	C1 C2	118.1(4)
C2 O3	122.6(6)		

The all *Zn*-O bonds in the zinc tetrahedra have same length (1.997(3) Å), however, the O-*Zn*-O angles visiable deviates from tetrahedral and are in the range of 93.0(2) - 128.9(3)° (**Table 3**). The C-O bond in the coordinated carbonyl group is visible longer than in other, 2.271(7) and 2.240(7) Å, respectively. The *Sal*-ligand has a planar comfiguration: the dihedral angle between benzoic ring and COO-group is 5.4(5)°. The internal O-H...O hydrogen bond between *ortho* arranged HO- and COO- groups in the benzoic cycle, observed in the free salicylic acis, is saved.

As shown in **Fig. 2**, the complex I contains two differently polarity parts: one of that involving the − C(3)H-C(4)H-C(5)H-

C(6)H–fragment of the benzoic cycle is a lipophilic and second part involving the coordinated two water molecules as well as the carboxylate group is a hydrophilic. This circumstance is very importance for the skin penetrating of the skin care ointment.

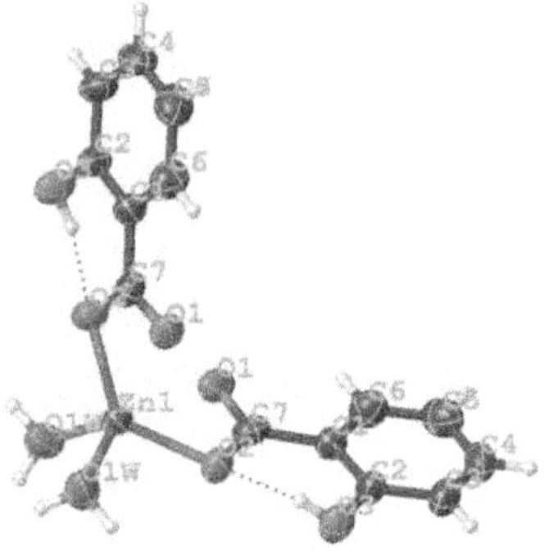

Fig.2. The molecular structure of I

The symmetry independent part of the complex I form two intermolecular O-H..O hydrogen bondsas the *p*-acceptor: first with OH-group of one neighbour molecule and second with the O atom of non- coordinated of the carbonyl group other molecule (**Figure 3a**). Intramolecular hydrogen bonding between phenolic and carboxylate groups occurs in many salicylic acid complexes [43].

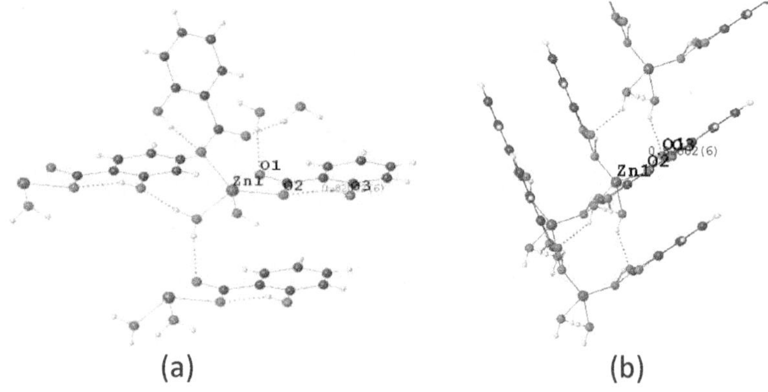

(a)   (b)

Figure 3. Fragment of the crystal packing of structure I. Dotted line is hydrogen bonds.

Fig.3 shows that in the complex I the benzoic cycles of the *Sal* ligands are oriented mutually almost perpendicularly. According **Fig.3b**, the benzoic cycles of the *Sal*-ligands of the two neighbor complexes are arranged coplanar one to other that is ensured minimal mutual attractions between cycles.

2.1.4 Conclusion

Thus, during the reaction of the zinc complex formation with salicylic acid, a complex compound of the composition $[Zn(Sal)_2(H_2O)_2]$ is formed (where the symbol *Sal* is the single charged anion of salicylic acid. The *Zn* atom is located on the symmetry axis 2 and has a tetrahedral coordination environment due to atoms O of two symmetrically related *Sal* anions and O atoms of two symmetrically related H2O molecules. The *Sal* anions coordinate the *Zn* atom in a monodentate *syn-1a* manner. The coordination tetrahedron has distortions in the O-Zn-O bond angles, which lie in the range of 109-120°. Molecule I has spatially separated hydrophilic and hydrophobic atom groups. That spatial separation of the molecule's regions in- dicates a high permeability of the molecule through the skin, which is important for creating an ointment based on the studied compound.

2.2 SYNTHESIS AND CRYSTAL STRUCTURE OF HEXAAQUANICKEL(II) BIS(4-HYDROXY- BENZOATE) DIHYDRATE

2.2.1 Chemical contest

The title compound, $[Ni(H_2O)_6](PHB)_2 \cdot 2H_2O$ (1) (PHB= 4-hydroxybenzoate, $C_7H_5O_3$), was obtained by the reaction of $NiCl_2$, 4-hydroxybenzoic acid (PHBA) and monoethanolamine in aqueous ethanol solution. The Ni^{II} ion is coordinated by six water molecules and is located on an inversion center. The outer coordination sphere in the asymmetric unit comprises one PHB anion and one water molecule, *i.e.* the compound is a salt and a hydrate consisting of three components. In the crystal, the components are packed into an intermolecular network stabilized by O—H O hydrogen bonds.

It can be anticipated that mixing a Brønsted base (MEA) with a Brønsted acid (PHBA) in a reaction medium also containing a metal salt ($NiCl_2$) may lead to the formation of different types of compounds: (*a*) the desired mixed-ligand Ni complex with MEA in neutral and PHBA in carboxylate forms; (*b*) both ligands coordinated in a neutral form with chlorine ions residing in the outer coordination sphere for compensation of the positive charge of the central nickel ion; (*c*) homoleptic complexes or those with only one organic ligand type plus water of coordination (and with or without anions in the outer coordination sphere for potentially needed charge compensation); or (*d*) a strictly organic salt between monoethanolammonium (*i.e.* protonated amine) and *para*-hydroxybenzoate (*i.e.* deprotonated acid, PHB). However, we have obtained (*e*), a supramolecular complex (**Fig. 4**) based on the Ni^{II} ion with six coordinated water molecules, two *para*-hydroxybenzoate anions in the outer coordination sphere and two lattice solvent water molecules.

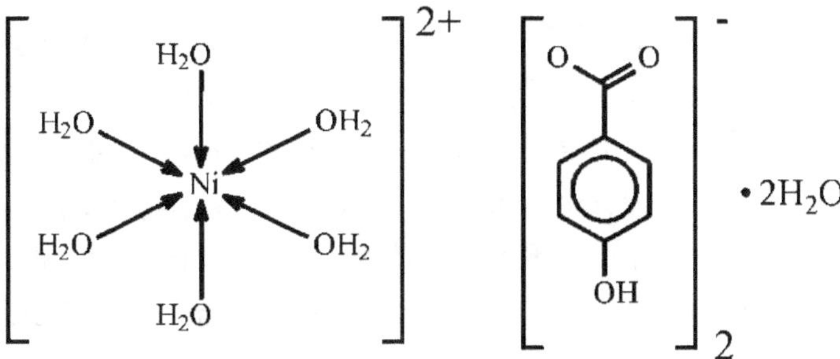

Fig. 4. Supramolecular complex based on the Ni^{II} ion with six water molecules, two *para*-hydroxybenzoate anions

We presume that this structure is realized due to the energetic favorability of the obtained complex, in particular in the solid state, since the formation of the hexaaquanickel(II) cation opens up the possibility of generating a multitude of stabilizing intermolecular hydrogen bonds. The Brønsted acid–base reaction between the two molecules intended as ligands apparently precedes complexation and/or crystallization and monoethanolamine or its protonated cationic ammonium form are absent from the crystallized salt. Nearly half a century ago, complexes of magnesium(II), cobalt(II) and manganese(II), which are isostructural to the compound reported here, were obtained and structurally characterized by a group in Azerbaijan (Shnulin *et al.*, 1981, 1984). The accuracy of these structure determinations was low, although reasonable for that time. An analogous nickel(II) complex with *p*-nitrobenzoate counter-ions was recently obtained and published by us [44]. Neither the intermolecular interactions of this analogous complex salt nor those in the isostructural compounds have been estimated quantitatively as yet. Notably, despite a search of the CSD (Groom *et al.*, 2016) for the hexaaquanickel(II) complex returning 352 hits, for only one of the reported crystal structures of $[Ni(H_2O)_6]^{2+}$ salts was a Hirshfeld surface analysis carried out [45]. This left the cationic complex unconsidered and a corresponding analysis of $[Ni(H_2O)_6]^{2+}$ is therefore unaccounted for to date. This communication is, hence, devoted to the crystal structure andcomprehensive Hirshfeld surface analysis of the obtained supramolecular complex salt 1.

2.2.2 Structural commentary

The molecular structure of 1 is show in **Fig. 5.** The asymmetric unit of the structure consists of half of the nickel complex ion (residing on an inversion center), one *para*-hydroxybenzoate anion (PHB) and one water molecule. The formula of the obtained compound is therefore $[Ni(H_2O)_6](PHB)_2\ 2H_2O$. The bond lengths between the metal center and the oxygen donor atoms of the water molecules fall into the small range 2.0483 (13)–2.0893 (13) $\mathring{A}$, while the bond angles vary between 88.72 (7) and 91.28 (7)°, *i.e.* the polyhedron around the central ion takes on the form of a nearly ideal octahedron. Compensation for the positive charge of the Ni^{++} ion is achieved with the deprotonation of PHBA molecules during the course of the reaction resulting in the respective carboxylate anions, which are incorporated in the outer coordination sphere. The carboxylate group is nearly but not perfectly coplanar with the aromatic ring evidenced by the corresponding dihedral angle of 12.51 (3)°. The complex cations interact with the anions through the formation of O7—H7$B\cdots$O1^{vi} [2.675 (2) Å] and O5—H5$B\cdots$O1 [2.632 (2) Å] hydrogen bonds (**Table 4**) with an $R^1(6)$ graph-set notation (Etter *et al.*, 1990).

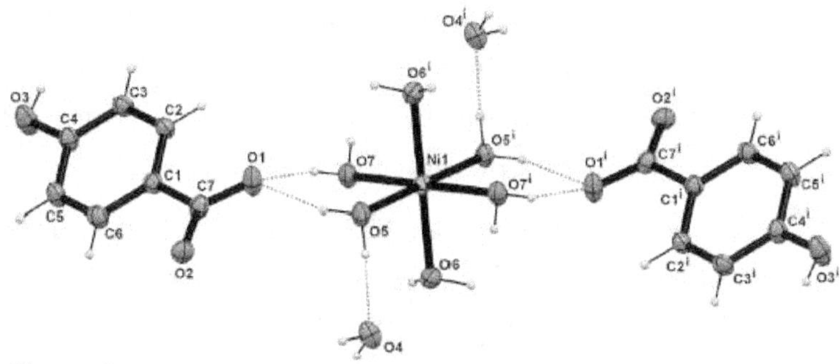

Fig 5. The molecular structure of 1. The ellipsoids of non-hydrogen atoms are drawn at the 50% probability level. Symmetry code: $1 - x, 1 - y, 1 - z$.

Table 4 Hydrogen-bond geometry (Å, °)

D—H$\cdots$$A$	D—H	H$\cdots$$A$	$D\cdots A$	D—H$\cdots$$A$
O3—H3$\cdots$O4	0.86 (3)	1.84 (3)	2.655 (2)	157 (3)
O4—H4$A\cdots$O5^{i}	0.82 (4)	1.97 (4)	2.785 (2)	173 (4)
O4—H4$B\cdots$O2^{ii}	0.87 (3)	1.88 (3)	2.724 (2)	163 (3)
O5—H5$A\cdots$O4^{iii}	0.87 (3)	1.91 (3)	2.770 (2)	169 (3)
O5—H5$B\cdots$O1	0.92 (4)	1.73 (4)	2.632 (2)	168 (3)
O6—H6$A\cdots$O2^{iv}	0.81 (3)	1.97 (3)	2.779 (2)	174 (3)
O6—H6$B\cdots$O2^{v}	0.79 (4)	2.00 (4)	2.748 (2)	157 (4)
O7—H7$A\cdots$O3^{iii}	0.84 (3)	1.88 (3)	2.723 (2)	176 (3)
O7—H7$B\cdots$O1^{vi}	0.93 (4)	1.79 (4)	2.675 (2)	159 (3)

Symmetry codes:

(i) $x, -y + \frac{3}{2}, z + \frac{1}{2}$; (ii) $x + \frac{1}{2}, y, -z + \frac{3}{2}$; (iii) $-x + \frac{1}{2}, -y + 1, z - \frac{1}{2}$; (iv) $-x + \frac{1}{2}, y - \frac{1}{2}, z$; (v) .

2.2.3 Supramolecular features

There are seven crystallographically independent oxygen atoms in the crystal structure, two of which serve only as hydrogen-bond acceptors (O1 and O2), three are both hydrogen-bond donors and acceptors (O3, O4, O5), and two are only hydrogen-bond donors (O6 and O7). All of the oxygen atoms are involved in relatively short intermolecular hydrogen bonds between the $[Ni(H_2O)_6]^{2+}$ cations, the PHB anions and the solvent water molecules. The $D \cdots A$ distances of these bonds are in the range 2.632 (2)–2.785 (2) Å (**Table 4**), which is indicative of sufficiently strong intermolecular interactions. The aromatic rings of the PHB anions are arranged in two different angles relative to the cell parameters and with an angle of 57.15° between their respective planes (**Fig. 6**).

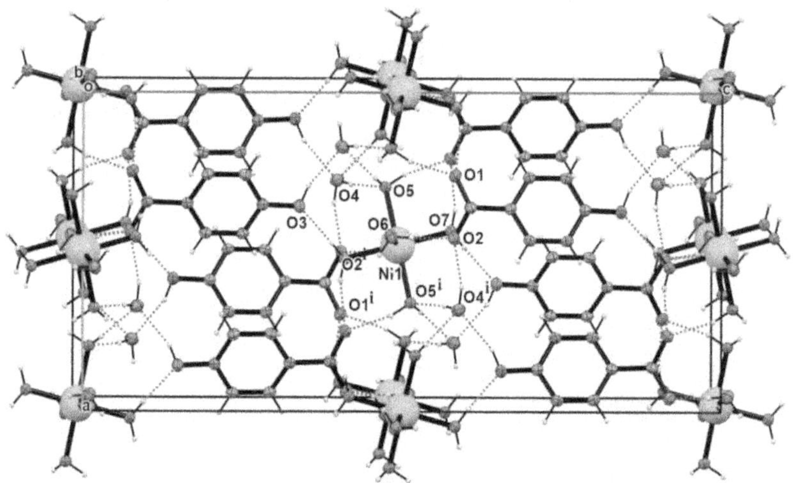

Fig. 6. The packing of 1 viewed along the *b*-axis direction.

Adjacent anions with the same ring alignment adopt opposite orientations (alcohol and carboxylate moieties on opposite sites of the molecules alternate when viewed along the crys- tallographic *a*-axis). The complex cations are bridged by the length of the 4-hydroxybenzoate anions in the *c*-axis direction. The cations are linked in the *ab* plane by hydrogen bonds to water molecules and the PHB alcohol and carboxylate moieties. In consequence, layers of organic and inorganic sublattices alternate in the *c*-axis direction. Together, these interactions associate the components into a three-dimen sional network (**Fig. 6**).

2.2.4 Database survey

A survey of the Cambridge Structural Database [Groom *et al.*, 2016; accessed January 2022 using *ConQuest* (Bruno *et al.*, 2002)] reveals that there are 352 hits in the database containing the hexaaquanickel(II) complex ion. Nearly half a century ago, coordination complex formation with benzoic acid derivatives including PHBA was widely studied in the Azerbaijan Institute of Applied Physics. Researchers from this institute synthesized and structurally characterized supra-molecular complexes analogous to compound 1 with magne-sium(II) (MGHBZA20; Shnulin *et al.*, 1981), cobalt(II) (MGHBZB20; Shnulin *et al.*, 1981) and manganese(II) (COLWUV; Shnulin *et al.*, 1984), which are all isostructural with the title compound. In addition, the structure of the View of the three-dimensional Hirshfeld surfaces for (*a*) the PHB anion and (*b*) the $[Ni(H_2O)_6]^{2+}$ cation of the title compound 1 plotted over d_{norm} in the range -0.4180 to 1.3344 a.u. magnesium(II) complex (AYOJOP; Baruah, 2016) is isomorphic with that of compound 1. The precision of the previous structure determinations of these compounds were not nearly as high as that of the structure reported here (*R*-factors of 0.07 or more compared to 0.03) while the inter-molecular interactions have not yet been assessed quantitatively.

2.2.5 Synthesis and crystallization

$NiCl_2$ (0.130 g, 1.0 mmol) was dissolved in a small amount of water. 4-Hydroxybenzoic acid (0.276 g, 2 mmol) was dissolved in a mixed solvent of 2 ml of absolute alcohol and 2 ml of distilled water. After dropwise addition of the PHBA solution and MEA to the nickel salt solution, the color changed gradually to light green. The resultant solution was stirred for 1 h with a magnetic stirrer at 318 K. The solution was allowedto stand at room temperature in a beaker with small holes in the cover for evaporation. About three weeks later, rectan- gular block-shaped single crystals of $[Ni(H_2O)_6](PH-BA)_2(H_2O)_2$ appeared. Analysis calculated: $NiC_{12}H_{26}O_{12}$: C, 34.22%; H, 6.18%. Found: C, 33.63%; H, 6.25%.

2.2.6 Structure refinement

Crystal data, data collection and structure refinement details for the structure of compound 1 are summarized in **Table 5**. The hydrogen atoms of water molecules and the hydroxyl group of the PHB anion were located in difference-Fourier maps and refined freely. The H atoms of the benzene ring were calculated geometrically with C—H = 0.93 Å and $U_{iso}(H) = 1.2U_{eq}(C)$.

Table 6. Refinement details for the structure

Chemical formula	$[Ni(H_2O)_6](C_7H_5O_3)_2 \cdot 2H_2O$
M_r	477.06
Crystal system, space group	Orthorhombic, *Pbca*
Temperature (K)	293
a, b, c (Å)	11.0812 (2), 7.63258 (17), 23.7986 (5)
V (Å^3)	2012.84 (7)
Z	4
Radiation type	Cu $K\alpha$
μ (mm^{-1})	2.05
Crystal size (mm)	0.2 ~ 0.18 ~ 0.15
Data collection Diffractometer	XtaLAB Synergy, single source at home/near, HyPix3000
Absorption correction	2020)
T_{min}, T_{max}	0.362, 1.000
No. of measured, independent and observed	$[I > 2\sigma(I)]$ reflections
R_{int}	0.033
$(\sin\theta/\lambda)_{max}$ (Å^{-1})	0.615
Refinement	$R[F^2 > 2\sigma(F^2)]$, $wR(F^2)$, S 0.034, 0.101, 1.05
No. of reflections	1949
No. of parameters	169
H-atom treatment	H atoms treated by a mixture of independent and constrained refinement
Δp_{max}, Δp_{min} (e Å^{-3})	0.25, −0.43

2.3 SYNTHESIS, CRYSTAL STRUCTURE OF THE Cu MIXED-LIGAND COMPLEX WITH 4-HYDROXYBENZOIC ACID AND MONOETHANOLAMINE

2.3.1 Introduction

The recently synthesized coordination compound [Cu(PHBA)$_2$(MEA)$_2$] (MC), where PHBA stands for p-hydroxybenzoic acid and MEA represents monoethanolamine, was obtained at a temperature of 28°C. The compound was formed by reacting PHBA, MEA, and freshly prepared copper hydroxide in an ethanol solution. The structure of MC was determined using single crystal X-ray diffraction, revealing that the complex molecules are positioned on inversion centers. The metal ions coordinate with two PHBA and two MEA molecules through both mono- and bidentate interactions, resulting in the formation of distorted octahedral polyhedra. The molecular arrangement is stabilized by strong intramolecular hydrogen bonds between the hydroxyl group of MEA and the oxygen atom of PHBA. The complex molecules are organized into a two-dimensional supramolecular substructure through intermolecular hydrogen bonds, which further assemble into a three-dimensional network through $\pi\cdots\pi$ interactions.

2.3.2 Synthesis and crystallization

The compound described in the title was synthesized using a two-step reaction. In the first step, copper hydroxide (1 mmole) was prepared by precipitating $CuCl_2$ with sodium hydroxide. In the second step, the resulting precipitate was dissolved by stirring while slowly adding monoethanolamine (MEA) drop by drop until a clear solution was obtained. The solution was then filtered, and an ethanol solution of p-hydroxybenzoic acid (PHBA) (2 mmole) was added to the filtrate at room temperature. The mixture was continuously stirred for 1 hour. The resulting solution, which was clear and green in color, was transferred into an evaporating dish with small holes and placed in a thermostat at 28°C. After two weeks, dark-green crystals formed. The crystals were collected, washed with acetone, and dried at room temperature.

The compound exhibited a yield of 62% during the synthesis. Elemental analysis was conducted for the compound with the molecular formula $C_{18}H_{24}CuN_2O_8$ (molecular weight: 459.93 g/mol). Calcd. C 22.62; H 5.26; N 6.09%; found C 22.05; H 5.15; N 5.90%.

2.3.3 X-ray structure analysis and refinement

Data for the crystal structure determinations were collected on an Oxford Diffraction Xcalibur-R CCD diffractometer (CuKa-radiation, λ 1.54184 Å, ω-scan mode, graphite monochromator) at 293 K. The structure was solved and refined using program packages SHELXT [46]a and SHELXL [46]b, respectively. All non-hydrogen atoms were refined anisotropically. Hydrogen atoms were inserted at calculated positions and constrained with isotropic thermal parameters. The molecular drawings were plotted using MERCURY program package [44]. The main crystallographic data and the structure refinement details are given in **Table 7**. Crystallographic data have been deposited with Cambridge Crystallographic Data center (Deposite Number 2039620). The data can be obtained free of charge via http://www.ccdc.cam.ac.uk/conts/ retrieving.html or from the Cambridge Crystallographic Data center, 12 Union Road, Cambridge CB2 1EZ, UK; fax: (+44) 1223-336-033; or e-mail: deposit@ccdc.cam.ac.uk.

Table 7. Main crystallographic data and the structure refinement details

Chemical formula	$C_{18} H_{24} Cu N_2 O_8$
M_r	459.93
Crystal system, space group	Triclinic, P-1
Temperature (K)	293
a, b, c (Å)	6.2344(6), 8.2288(5), 10.3573(9)
V (Å^3)	501.07(7)
Z	1
Radiation type	Cu $K\alpha$
μ (mm^{-1})	1.972
Crystal size (mm)	0.15 x 0.22 x 0.25
Data collection Diffractometer	XtaLAB Synergy, single source at home/near, HyPix3000
Absorption correction	Multi-scan (*CrysAlis PRO*; Rigaku OD, 2020)
T_{min}, T_{max}	0.362, 1.000
[$I > 2\sigma(I)$] reflections	0.0402, 0.1077
R_{int}	0.033
$(\sin \theta/\lambda)_{max}$ (Å^{-1})	0.615
Δp_{max}, Δp_{min} (e Å^{-3})	-0.50/0.47

2.3.4 Description of molecular structure

The complex molecule is situated on an inversion center of the triclinic P-1 system. In it Cu(II) ion coordinates two PHBA and two MEA molecules, *i.e.* formula of the coordination compound is [Cu(PHBA)$_2$(MEA)$_2$] (**Fig. 7**). The PHBA molecules are attached to metal ion monodentately through oxygen atom O1 of the carboxylic group while MEA molecules are bound by chelate fashion through N1 and O4 atoms.

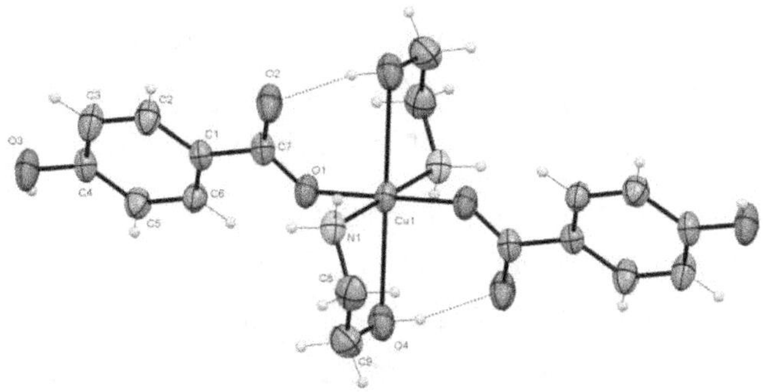

Fig. 7. The ORTEP structure of the MC. An asymmetric part of the complex molecule is numbered and ellipsoids are drawn at 50% probability level.

Polyhedron of the central atom is octahedron which is distinctly distorted (**Fig. 7**). The distortion is mainly caused by the pro- longation of the apical bonds until 2.609 Å due to Jahn-Teller effect. Indeed, four equatorial bonds of a type Cu-O and Cu-N are equal (1.991(2)) Å and angles with participation of these atoms are 90 1.3° whereas angles based on apical atoms O4 are deviated from the ideal value until 15° (**Table 8**).

Table 8 Bond distances and angles of the Cu atom.

Bond	Value, Å
Cu1-O1	1.992(2)
Cu1-O4	2.609(2)
Cu1-N1	1.991(2)
Angle	**Value, °**
O1-Cu1-O4	88.71(7)
O1-Cu1-N1	90.28(8)
O1-Cu1-O4^i	91.29(7)
O1-Cu1-N1^i	89.72(8)
O4-Cu1-N1	75.03(8)
O4-Cu1-N1^i	104.97(8)

Symmetry code: (i) 1-x,1-y, -z.

In order to compensate positive charge of the Cu(II) ion the carboxyl group of PHBA is deprotonated giving rise to its carboxylate form. The carboxylate fragment is tilted to aromatic ring at an angle of 10.92°. As the result of the bidentate coordination of the MEA molecule five-membered ring is formed. The atoms of this ring are not coplanar but its conformation resembles a half-chair form.

The complex molecule is characterized by strong intramolecular H-bond O4-H···O2 (**Table 9**). This H-bond closes 6-membered ring with a graph-set notation of $S^1_1(6)$ [47].

Table 9. The hydrogen bonds in the structure of MC.

Bond, D–H···A	D–H, Å	H···A, Å	D···A, Å	∠D–H···A, °
O4-H4···O2^{ii}	0.8200	1.8100	2.605(3)	163.00
O3-H3···O4^{i}	0.8200	1.8700	2.675(3)	169.00
N1-H1A···O3^{iii}	0.8900	2.2500	3.073(3)	153.00
N1-H1B···O2^{iv}	0.8900	2.1000	2.929(3)	154.00

Sym. codes: (i) 1-x, 1-y, 1-z; (ii) 1-x, 1-y, -z; (iii) 2-x, 1-y, 1-z; (iv) 2-x,1-y,-z.

2.3.5 Supramolecular features and hydrogen bonds

The crystal structure consists of three proton-donor groups (O3-H, O4-H, and N-H2) and two proton-acceptor groups (O1 and O2). With the exception of the O1 atom, all of these groups fulfill their respective roles as proton donors or acceptors. Additionally, both O3-H and O4-H hydroxyl groups act as proton acceptors, forming single hydrogen bonds. Overall, the crystal structure is defined by the presence of three intermolecular hydrogen bonds. The H-bond O3-H3··· O4 incorporates of complex molecules in the direction of *c*-

axis, N1-H1B   O2 – in the direction of *a*-axis, while N1-H1A ···O3 – in the diagonal direction (**Table 9**). These H-bonds incorporate molecules into 2-D associates parallel to (010) plane. Further these associates are incorporated into 3-D network by π···π interactions for which a distance between phenyl rings is 3.702 Å (**Fig. 8**).

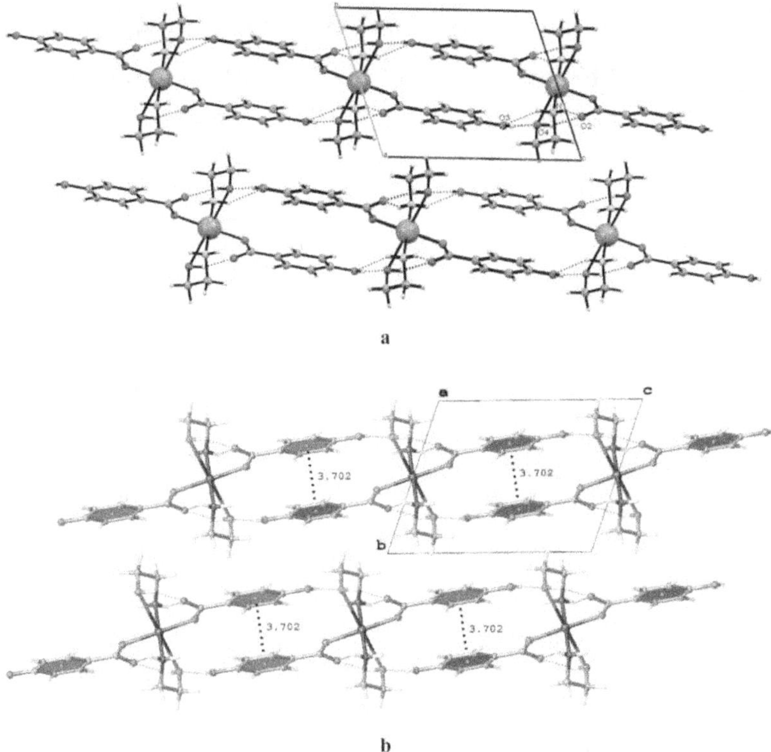

Fig. 8. Crystal structure of MC: (a) intermolecular H-bonds and (b) π···π interactions.

2.3.6 Database survey

There are 528 structures with participation of PHBA in Cambridge Structural Database (CSD, version 5.42, November 2020, updated on February, 2021)) [38]. Among them 183 structures are metal complexes. There is only one none-metal complex (organic salt) on the basis of PHBA and MEA with ref-code KEZFIG [48]. The only mixed-ligand metal complex of PHBA and MEA is Cd-complex of these ligands with ref-code YEBSIK [49]. In this complex Cd(II) ion bidentately coordinates two PHBA and two MEA molecules thus having a coordination number eight. Unfortunately, a paper [49] is devoted only to the X-ray structure of the Cd-complex without testing its bioactivity.

PART 3. NOVEL BINUCLEAR COPPER COMPLEXES WITH SA/PHBA LIGANDS

3.1 INTRODUCTION

Transition metal complexes containing ortho- and para-hydroxybenzoic acids have become highly significant in coordination chemistry, crystal engineering, and potential applications. These complexes have garnered significant interest due to their distinct properties and wide-ranging potential in diverse fields. [50]. Therefore, metal complexes containing such type ligands are frequently studied by means of spectroscopic, structural and thermal methods [51–53]. Carboxylates have been found to display a variety of bonding modes, including monodentate, monodentate bridging, symmetric and asymmetric chelating, and bidentate arrangements. This flexibility in bonding modes contributes to the broad range of coordination chemistry observed in metal complexes with carboxylate ligands. [54, 55]. Generally, a bulky benzene ring substituent in an ortho/para position to the carboxylato group is directly bonded with Cu(II) and resulting in the formation of dinuclear paddle-wheel like Cu(II) complexes [54–56]. In this particular case, the crystal structures of the Cu(II) complex revealed a dimeric paddle-wheel cage structure. Within this structure, the copper ions were bonded to oxygen atoms in both equatorial and axial positions, forming Cu-O bonds.

Recently, the significance of intra- and intermolecular hydrogen bonding interactions in crystalline Cu(II) dinuclear complexes has been highlighted for achieving solid-state square pyramidal structures. Extensive endeavors have been dedicated to designing metal complex structures by utilizing hydrogen bonding and π-π stacking interactions. This involves a meticulous selection of building blocks and organic ligands in the field of supramolecular chemistry. [56–60]. Developing a novel strategy based on hydrogen bonding for crystal engineering poses significant challenges. However, this research work represents a continuation of our previous studies focused on the synthesis and properties of metal complexes incorporating derivatives of benzoic acids. By exploring the potential of hydrogen bonding interactions, we aim to advance our understanding of crystal engineering principles and expand the possibilities for designing and controlling the structures and properties of metal complexes. [60]. Indeed, our research is committed to exploring binuclear Cu(II) complexes of *ortho* and *para* hydroxybenzoic acid as O donor ligands. Here each *o*-hydroxybenzoic/*p*-hydroxybenzoic acid served as a monobasic but bidentate ligand.

The new paddlewheel-like dinuclear copper complexes of general formulae $[Cu_2(o\text{-}HBA)_4(DMFA)_2]$ (1), $[Cu_2(o\text{-}HBA)_4(H_2O)_2]DMFA$ (2) and $\{[Cu_2(o\text{-}HBA)_2(p\text{-}HBA)_2(H_2O)_2]EtOH\}\{[Cu_2(o\text{-}HBA)_4(H_2O)_2]\}$, (3) (*o/p*-HBA=*ortho/para*-hydroxybenzoic acid, DMFA=Di-methylforamide) have been obtained by the reaction in ethanol, water and water : ethanol solution of MEA (monoethanolamine) and freshly prepared copper hydroxide with *o*-HBA (for 1 & 2) and *o/p*-HBA (for 3) respectively. In the investigated ligands, each one was coordinated to the metal ions through two oxygen atoms derived from the carboxylic acid group. Furthermore, in the investigated complexes, a solvent molecule, such as DMFA (N,N-dimethylformamide) or water, was attached to each Cu(II) atom at the axial position through oxygen atoms. In the case of Complex-3, a distinct approach was employed, where two dinuclear Cu(II) complexes were linked together through water molecules using hydrogen bonding interactions. This alternative approach led to the formation of a distinctive structure specific to Complex-3.

3.2 EXPERIMENTAL PART

3.2.1 Synthesis of $[Cu_2(o\text{-}HBA)_4(DMFA)_2]$ (complex-1)

To synthesize the metal complex depicted in Figure 9, a two-step reaction was employed. In the first step, CuSO4 was reacted with NaOH, resulting in the precipitation of a new copper hydroxide (1 mmole). Subsequently, in the second stage, the obtained precipitate was dissolved by stirring while gradually adding an ethanol solution of monoethanolamine (MEA) until a clear solution was formed. The resulting solution was then filtered, and a solution of p-hydroxybenzoic acid (PHBA) in N,N-dimethylformamide (DMFA) (2 mmole) was added to the filtrate while continuously stirring at room temperature for 1 hour. The resulting solution, which exhibited a clear green color, was transferred to an evaporating glass container with small holes and placed in a thermostat set at 28 degrees Celsius. After a period of two weeks, dark green crystals formed in the solution. These crystals were collected, treated with acetone for washing, and subsequently dried at room temperature in the air. Yield: 51%. Elemental analysis for $C_{34}H_{34}N_2O_{14}Cu_2$(821.73): Cal. C, 49.56; H 4.13; N 3.4%; found C 49.01; H4.09; N 4.09%.

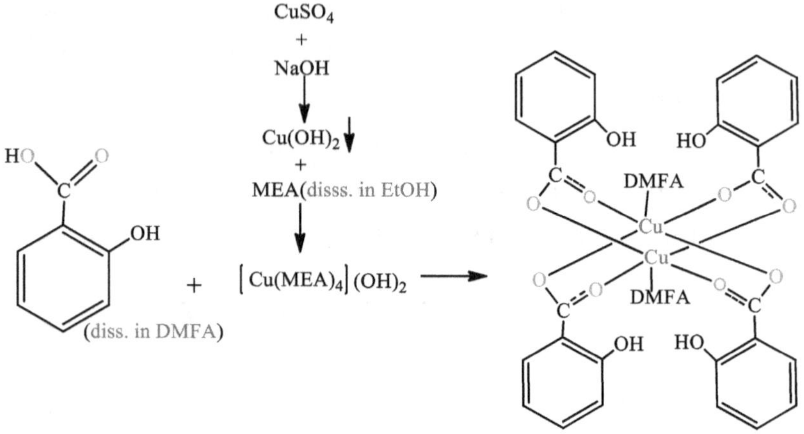

Fig.9. Preparation of [Cu₂(*o*-HBA)₄(DMFA)₂], complex-1

3.2.2 Synthesis of [Cu₂(*o*-HBA)₄(H₂O)₂]DMFA (complex- 2)

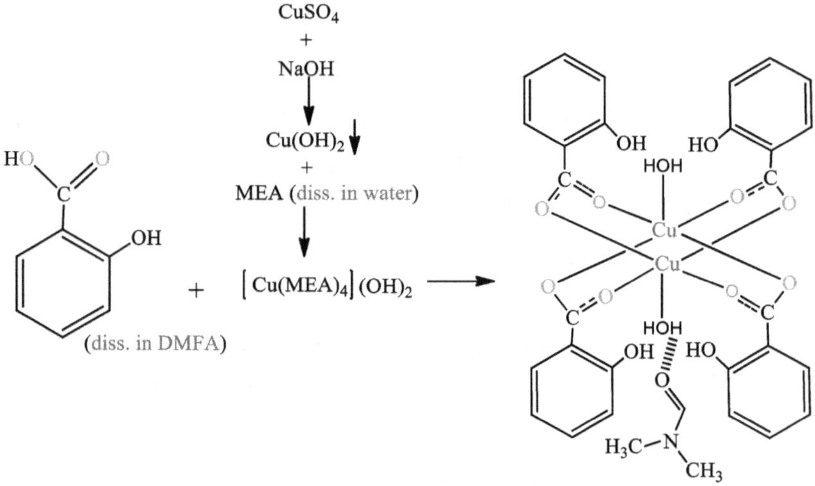

Fig.10. Preparation of
[Cu₂(*o*-HBA)₄(H₂O)₂]DMFA,complex-2

The complex-2, with the general formula [Cu2(o-HBA)4(H2O)2]DMFA, was synthesized using a similar method as complex-1 (as shown in Figure 10), but with a different solvent mixture. Instead of using an EtOH+MEA mixture, an H2O+MEA mixture was employed. The yield for the synthesis of complex-2 was 60%. Elemental analysis was conducted for the compound with the molecular formula $C_{31}H_{31}NO_{15}Cu_2$ (molecular weight: 784.67 g/mol). The calculated percentage composition was found to be: C 49.68%; H 4.14%; N 1.87%. The experimental analysis yielded slightly different results with the following composition: C 49.57%; H 4.08%; N 1.85%.

3.2.3 Synthesis of {[Cu₂(*o*-HBA)₂(*p*-HBA)₂(H₂O)₂]EtOH}{[Cu₂(*o*-HBA)₄(H₂O)₂]} (complex-3):

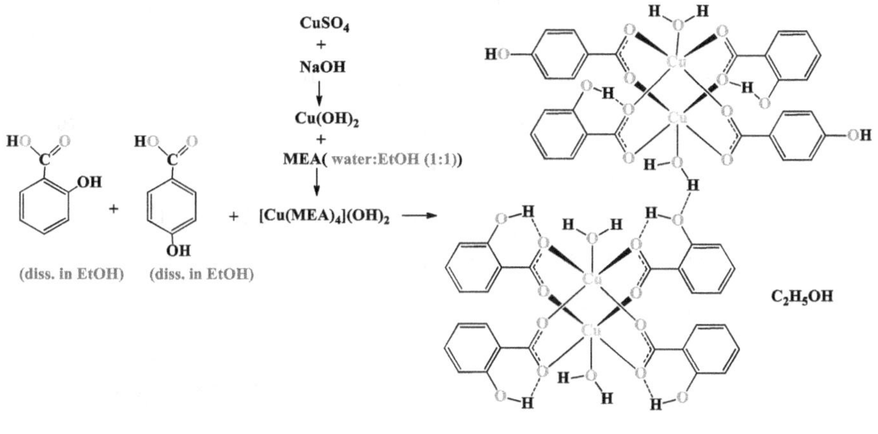

Fig. 11. Preparation of {[Cu₂(SA)₂(PHBA)₂(H₂O)₂]EtOH}{[Cu₂(SA)₄(H₂O)₂]}, complex-3

Complex-3 has been prepared a slightly different procedure (**Fig.11**). Freshly prepared copper (II) hydroxide (0.5 mmole) was dissolved in a water+ethanol solution of MEA. *o*-, and *p*-HBA (1 mmole of each ligand) were dissolved separately at room temperature in ethanol. Initially, o-HBA and then *p*-

HBA solution was added dropwise to the clear dark blue copper metal solution respectively. After slowly evaporation of the resulting solution at room temperature the quality crystals were obtained after 25 days. Yield: 50% Elemental analysis for $C_{58}H_{54}O_{29}Cu_4$ (1469.21): Cal. C, 47.37; H, 3.675; N, 1.87; found C, 46.95; H, 3.570; N, 1.85%.

3.2.4 X-ray structure analysis and refinement

Data for the crystal structure determinations were collected on an Oxford Diffraction Xcalibur-R CCD diffractometer (CuKa-radiation, λ=1.54184 Å, ω-scan mode, graphite monochromator) at 293 K [61]. The structure was solved and refined using program packages SHELXT [61] and [61], respectively. All non-hydrogen atoms were refined anisotropically. Hydrogen atoms were inserted at calculated positions and constrained with isotropic thermal parameters. The molecular drawings were plotted using MERCURY program package[43].

3.3 RESULTS AND DISCUSSION

3.3.1 Description of the crystal structures of complex-1,2,3

The bond lengths and bond angles for complexes 1, 2 and 3, are included in **Table 10, 11** and **12**, respectively and crystallographic data and structure refinement summary for both complexes are included in **Table 4**. The hydrogen bonds in the structures of complex-1, 2 and 3 are listed in the supplementary file.

Structure of [Cu₂(*o*-HBA)₄(DMFA)₂], complex-1: The molecular structure and atomic numbering scheme is shown in **Fig. 10.** Orthorhombic crystal system has been recognized according to Crystal data.

The binuclear Cu(II) complex was unambiguously determined by a single-crystal X-ray analysis. Based on bond lengths and bond angles the structure has revealed as a distorted square pyramidal shape (**Fig. 10**).

Here, binuclear Cu(II) structure was formed by two coordination spheres each of which were filled by four oxygen atoms from COO⁻ group and fifth oxygen atom from DMFA solvent. Each DMAF molecule was coordinated through oxygen to the copper(II) atom [62].

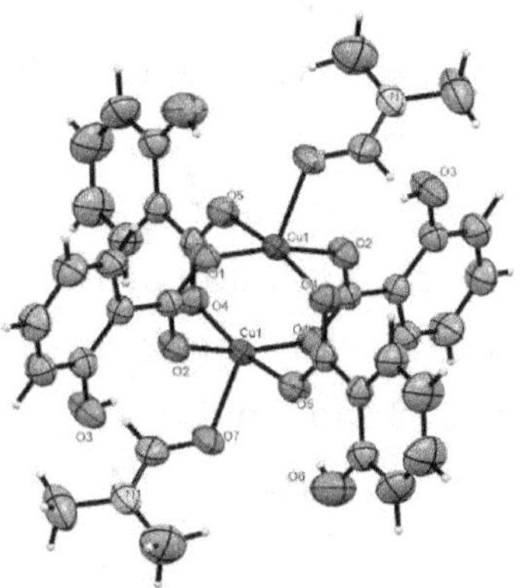

Fig 10. ORTEP diagram for the structure of [Cu₂(*o*-HBA)₄(DMFA)₂]

The complex has revealed a paddlewheel binuclear Cu(II) structure, in which each Cu(II) ion has bonded with four O atoms at the equatorial positions and one O atom of DMFA molecule was at the apical position (**Fig. 10**) [56, 63, 64]. The four equatorial bonds of the Cu-O type were nearly equal and the two axial Cu1-O7Å bonds were of the same. (**Table 10.**). The average equatorial Cu-O bond distance was 1.965(2) Å and the oxygen atom of DMFA, was lying at a distance of 2.134(2)

Å. These values are almost analogous with previous literatures [56, 65]. The copper atom deviates from the plane of the four para carboxylic oxygen atoms toward the DMFA molecule by 0.169Å. Here hydrogen bonds have formed through O3-H3···O2 and O6-H6A···O5 between ortho hydroxyl groups and para carboxylic group in same molecules. In the previous literatures the bridging was happened by the slight prolongation of the Cu-Cu bonds length up to 2.6433Å [56, 65]. But here Cu-Cu bridging was not observed however the structure was somewhat similar [56, 66, 67]. The bond angle O5–Cu1–O1was slightly deviated to O2-Cu1-O4 and was fully matched with related complexes [56, 68, 69]. The slightly distortion of the square pyramids was revealed by the bond angles between the apical and the equatorial donor atoms, which showed a deviation of 1.13 (O4[1]- Cu1- O2[1]) for comlex-1 and from the ideal 90° angle in a regular square pyramid.

Table 10. Bond lengths and bond angles of the [$Cu_2(o$-HBA)$_4$(DMFA)$_2$] complex

Bond	Length/Å
Cu1… Cu1[1]	2.6433(8)
Cu1- O4[1]	1.957(2)
Cu1- O5	1.961(2)
Cu1- O1	1.954(2)
Cu1- O2[1]	2.002(2)
Cu1- O7	2.134(2)
Angle	**Angle/°**
O4[1]- Cu1- O5	167.94(9)
O4[1]- Cu1- O2[1]	88.87(10)
O4[1]- Cu1- O7	98.48(10)

O5- Cu1- O2^1	88.46(10)
O5- Cu1- O7	93.31(10)
O1- Cu1- O4^1	89.45(10)
O1- Cu1- O5	90.76(10)
O1- Cu1- O2^1	168.18(9)

Symmetry codes: ^{1}1-x,1-y,1-z

Structure of [Cu₂(*o*-HBA)₄(H₂O)₂]DMFA, complex-2: In the crystal structure category, complex-2 has treated as a monoclinic crystal system (**Table 13**). The crystal was built as a binuclear Cu(II) complex along with two water molecules were bonded at axial position and also linked to dimethylformamide through hydrogen bond (**Fig.11**).

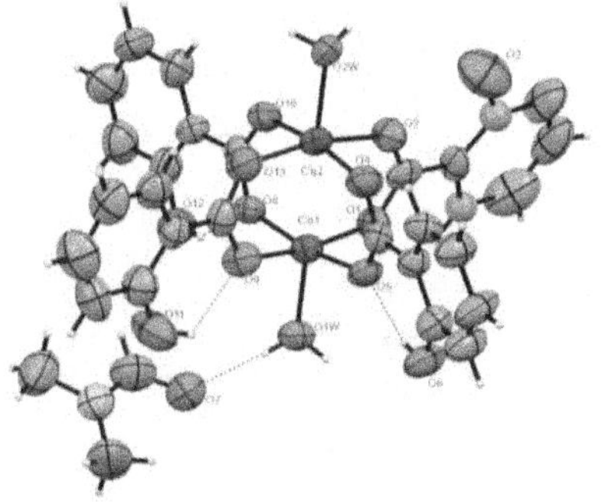

Fig 11: ORTEP diagram for the structure of [Cu₂(*o*-HBA)₄(H₂O)₂]DMFA, Complex-2

Here four equatorial bonds of a type Cu-O and axial Cu-O bond lengths were almost same like Complex -1. The metal — oxygen bond at axial and equatorial position were formed almost similar fashion. So based on different bond lengths the complex has been established as a distorted square pyramidal [56, 70]. The value of bond lengths and bond angles for complex-2 have revealed similar pattern like complex- 1. Here, according to ORTEP diagram, the intermolecular hydrogen bonds have found between H of the water molecules and O of the DMFA solvate molecules. The paddlewheel-like (waterwheel) binuclear Cu(II) structure has built in which each Cu(II) ion has bonded with four O atoms of different carboxylate groups in the equatorial and a water molecule at the axial positions, as a result the square pyramidal coordination geometry has formed. and the distance between the two atoms Cu2···Cu1 was found at 2.608(2)Å [56, 64, 71].

Table 11: Bond length and Bond angle of [Cu2(o-HBA)$_4$(H2O)$_2$]DMFA,Complex-2

Atom	Length/Å
Cu2...Cu1	2.608(2)
Cu2-O2	1.964(7)
Cu2-O2W	2.157(7)
Cu2-O4	1.947(7)
Cu2-O10	1.946(7)
Cu2-O13	1.959(8)
Angle	**Angle/°**
O9–Cu1–O8	88.3(3)
O5-Cu1-O1	89.9(3)
O4- Cu2- O2	89.6(3)
O4- Cu2- O2 W	96.6(3)
O4- Cu2- O13	87.3(3)
O13- Cu2- Cu1	84.6(3))
O13- Cu2- O2	168.5(3))
O10- Cu2- O2W	94.4(3)
O10- Cu2- O4	168.9(3)

The major bond angles O9–Cu1–O8, O5-Cu1-O1, O13-Cu2-O10 & O4-Cu2-O2 have exhibited 88.3(3)°,89.9(3)°, 89.7(3)° and 89.6(3)° respectively whose values were almost similar (**Table 11**) and these bond angles values were typically relevant of square pyramidal structure for binuclear Cu(II)complexes [56, 71]. The packing of molecules in crystal (complex-2) was primarily determined by the hydrogen bonds between the water molecules and the DMAF solvent molecules and complex-2 has characterized by strong intermolecular H-bond, O1W-H1WB⋯O7 [53, 56] (**Table 11**).

Structure of {[Cu$_2$(*SA*)$_2$(*PHBA*)$_2$(H$_2$O)$_2$]EtOH}{[Cu$_2$(*SA*)$_4$(H$_2$O)$_2$]}, complex-3: An ORTEP diagram of complex-3 with the corresponding atom numbering scheme is shown in **Figure 12.** The unit cell composed of two binuclear molecules and one C$_2$H$_5$OH group. Here two group of binuclear Cu (II) complexes were coordinated by hydrogen bond through water molecules. Each Cu(II) atom of one group of binuclear complex was bonded with water molecules at axial position like complex-2. Here C$_2$H$_5$OH molecules has not coordinated directly with the main body of the complex but acted as a solvent and play an important role in crystal structure organization.

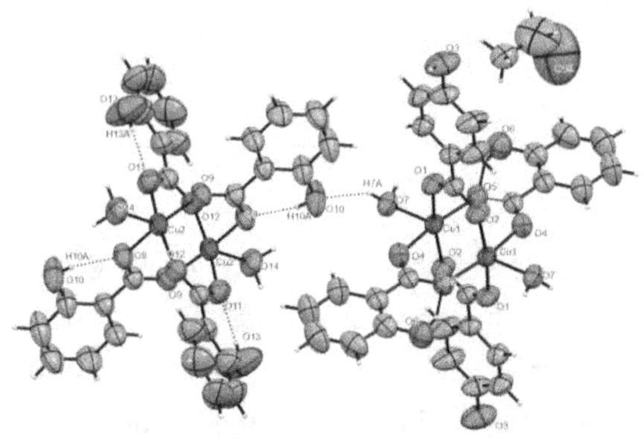

Fig 12. ORTEP diagram for the structure of
{[Cu$_2$(*o*-HBA)$_2$(*p*-HBA)$_2$(H$_2$O)$_2$]EtOH}{[Cu$_2$(*o*-HBA)$_4$(H$_2$O)$_2$]},
Complex- 3

Table 12. Bond length and Bond angle of $\{[Cu_2(o\text{-}HBA)_2(p\text{-}HBA)_2(H_2O)_2]EtOH\}\{[Cu_2(o\text{-}HBA)_4(H_2O)_2]\}$,Complex-3

Atom	Length/Å
Cu1..Cu1	2.6008(11)
Cu1-O5	1.976(3)
Cu1-O4	1.959(3)
Cu1-O2	1.955(3)
Cu1-O1	1.946(3)
Cu1-O7	2.162(3)
Cu2-O14	2.152(3)
Cu2...Cu2	2.6282(10)
Angle	**Angle/°**
O5-Cu1-Cu1[1]	86.88(9)
O4- Cu1- O7	98.13(13)
O4- Cu1- Cu1[1]	82.58(9)
O4- Cu1- O5[1]	169.45(13)
O7-Cu1-Cu1	172.65(9)
O2- Cu1-Cu1[1]	83.97(9))
O9-Cu2-O8[2]	168.08(12)
O14-Cu2-Cu2[2]	173.44(12

[1]-X,1-Y,-Z; [2]-X,2-Y,1-Z

The distances of Cu1...Cu1 and Cu2...Cu2 were found to be 2.6008Å and 2.6282Å from the crystal structure, respectively, were generally elongated compared to the other bond lengths and found to be similar to complexes-1 and 2.

The notable observation was that the axial bond lengths of O7-Cu1 Å and Cu2-O14 Å were almost longer than the equatorial bond lengths of the complexes. They were mostly identical to the distorted square pyramidal shaped complexes

[56, 62, 64]. In fact, the distortion was mainly caused by the prolongation of the axial bonds due to Jahn-Teller effect. In the complex, each binuclear Cu(II) complex have been followed the same sequence in terms of bond length and bond angle arrangement.

Indeed, there are numerous transition metal species with binuclear paddle wheel-like structural features, whose are a new area for research by coordination chemists to design functional metal–organic frameworks (MOFs) [72, 73]. But in this case the complex-3 was formed from ortho and para-hydroxy benzoic acid and eventually it has composed of two binuclear paddlewheel-like complexes linked by intermolecular hydrogen bond, and intra-molecular hydrogen bonds O–H...O have also appeared between H atom of ortho-hydroxy group and carboxylic oxygen atoms [56]. In addition, that strong inter (O7--H7A$\cdots$O10) and intramolecular (O10 -- H10A$\cdots$O8) hydrogen bonds were observed in the complex-3 [56][10].

Table 13. Crystallographic data and structure refinement summary for complex 1, 2 and 3

Empirical formula	$C_{34}H_{34}N_2O_{14}Cu_2$ (Complex-1)	$C_{31}H_{31}NO_{15}Cu_2$ (Complex-2)	$C_{58}H_{54}O_{29}Cu_4$ (Complex-3)
Formula weight	821.73	784.67	1469.21
Temperature/K	293 K	293K	293 K
Crystal system	Orthorhombic	Monoclinic	Triclinic
Space group	P b c a	P2$_1$/n	P-1
a/Å	17.0693(3)	10.4127(10)	10.3099(7)
b/Å	10.6543(2)	17.3020(16)	12.1640(8)
c/Å	19.3715(3	19.1787(18)	14.7468(6)
α/°	90	90	89.194(4)
β/°	90	104.619(9)	82.638(4)
γ/°	90	90	74.669(6)

			1768.50(19)
Volume/Å^3	3522.93(11)	3343.4(6)	
Z	15	15	8
ρ_{calc}g/cm^3	1.667	1.559	1.423
μ/mm^{-1}	2.956	2.211	3.140
F(000)	1800	1608	776
Radiation	CuKa(λ = 1.54184)	Cu Ka (λ = 1.54184)	CuKa(λ = 1.54184)
2Θ range for data collection/°	9.13 to 135.282	6.986 to 143.02	3.0 to 67.7
Index ranges	$-20 \leq h \leq 19$; $-12 \leq k \leq 12$; $-23 \leq l \leq 21$	$-10 \leq h \leq 12$; $-21 \leq k \leq 21$; $-23 \leq l \leq 23$	$-10 \leq h \leq 12$; $-14 \leq k \leq 14$; $-17 \leq l \leq 17$
Reflections collected	13617	17957	8792
Independent reflections	6020 [R_{int} = 0.03, R_{sigma} = 0.60]	6020 [R_{int} = 0.1984, R_{sigma} = 0.2648]	5182 [R_{int} = 0.0215, R_{sigma} = 0.0352]
Data/restraints/parameters	3163/0/239	6020/0/444	5182/0/433
Goodness-of-fit on F^2	1.057	0.887	1.06
Final R indexes [I>=2σ (I)]	R$_1$=0.0489, wR$_2$=0.15	R$_1$ = 0.0967, wR$_2$ = 0.2065	R$_1$ = 0.0549, wR$_2$ =0.165
Final R indexes [all data]	R$_1$ = 0.0556, wR$_2$ = 0.1731	R$_1$ = 0.2659, wR$_2$ = 0.2935	R$_1$ = 0.0663, wR$_2$ =0.177
Larg. diff. peak/hole/e Å^{-3}	1.834/-0.56	0.64/-0.58	0.64/-0.458

3.3.2 IR spectra analysis

IR spectroscopic analysis was conducted in order to define the three complexes. The positions of the bands connected with vibrations of the OH stretching, carboxylate and M-CO groups were analyzed in the range of 3500–3100 cm^{-1}, 1600–1300 cm^{-1} and 530-531 cm^{-1} respectively. Those Infrared bands of Cu$_2$(o-HBA)$_4$(DMFA)$_2$] complex-1, [Cu$_2$(o-HBA)$_4$(H$_2$O)$_2$]DMFA, complex-2 and {[Cu$_2$(o-HBA)$_2$(p-HBA)$_2$(H$_2$O)$_2$]EtOH}{[Cu$_2$(o-HBA)$_4$(H$_2$O)$_2$]}, complex-3 were assigned clearly due the presence of all characteristic functional groups. Specially, in the [Cu$_2$(o-HBA)$_4$(H$_2$O)$_2$]DMFA, complex-2 and {[Cu$_2$(o-HBA)$_2$(p-HBA)$_2$(H$_2$O)$_2$]EtOH}{[Cu$_2$(o-HBA)$_4$(H$_2$O)$_2$]}, complex-3, the broad and strong band between 3200 and 3700 cm^{-1} were attributed to the O–H stretching mode of water molecules [50, 70, 71]. Generally the narrow or broad absorptions bands are appeared between 1400-1300 cm^{-1} and 1700-1500 cm^{-1} for v_s (COO$^-$) and v_{as} (COO$^-$) respectively for stretching mode of vibration of carboxylic group [74, 75]. Here strong bands have appeared of the three metal complexes structures between 1605 to 1655 cm^{-1} which have clearly ascribed due to v_{as} (COO$^-$) vibrations of the carboxylate groups of the o-hydroxybenzoate and p-hydroxybenzoate anions. However, the strong bands have observed between 1383-1385 cm^{-1} for complex 1, 2 and 3, due to the v_s (COO$^-$) vibrations [62, 63, 70, 71, 74–77]. The v(C–H) stretching vibrations was found in a very weak absorption at 2929, 3002 cm^{-1} for complex-1 and 3001 cm^{-1} for complex-2, and but unfortunately this type of absorption peak did not observe for comples-3 due to overlap or unknown reason. The COH bonds of o/p-hydroxybenzoate ion was free because strong v(COH) bands appear within 1207-1244 cm^{-1} and these bending vibrations were fully detectable for three complexes [77–79]. Three binuclear complexes, the phenolic bands were found at 1236–1244 cm^{-1} [75]. Moreover, big bands have found within the range of 530-531 cm^{-1} of the complexes due to M-CO bonds [80].

3.3.3. UV-vis absorption spectra

The UV–Vis absorption spectrum of the three complexes in DMSO medium in the region around 200–970 nm and below 300nm bands are attributed due to π-π* transition of the aromatic rings [80]. All complexes displayed mainly three bands in the region 347 nm (28800 cm^{-1}), 638 nm (15680 cm^{-1}), and at 900nm (11100 cm^{-1}). These absorption bands were observed due to d-d transitions for binuclear complexes [70, 81].

PART 4. APPLYING HIRSHFELD SURFACE ANALYSIS TO INVESTIGATE MOLECULAR INTERACTIONS IN METAL COMPLEXES OF SA/PHBA

4.1 HIRSHFELD SURFACE ANALYSIS OF [Ni(H$_2$O)$_6$](PHB)$_2$·2H$_2$O

Intermolecular interactions can be assessed quantitatively by carrying out a Hirshfeld surface analysis [82]. We have calculated Hirshfeld surfaces and fingerprint plots separately for the PHB anion and [Ni(H$_2$O)$_6$]$^{2+}$ cation of compound 1. The red spots on the surfaces show the predominant strong interactions, which correspond to the O6—H6A O2, O3—H3 O4, O5—H5A O1, O7—H7B O3, and O7—H7A ··O1 hydrogen bonds, whereas the blue areas represent regions completely free from close contacts (**Fig. 13**).

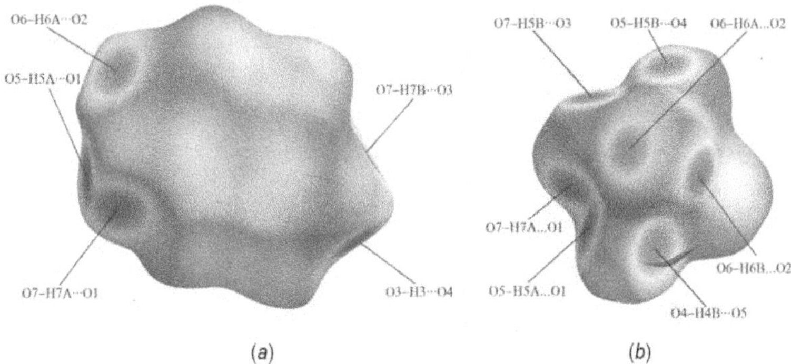

Figure 13. View of the three-dimensional Hirshfeld surfaces for (*a*) the PHB anion and (*b*) the [Ni(H$_2$O)$_6$]$^{2+}$ cation of the title compound 1 plotted over d_{norm} in the range —0.4180 to 1.3344 a.u.

Despite the high molecular symmetry of the complex cation, there are differences with regard to its Hirshfeld surfaces between the aqua ligands. Two aqua ligands (O5, O5A,

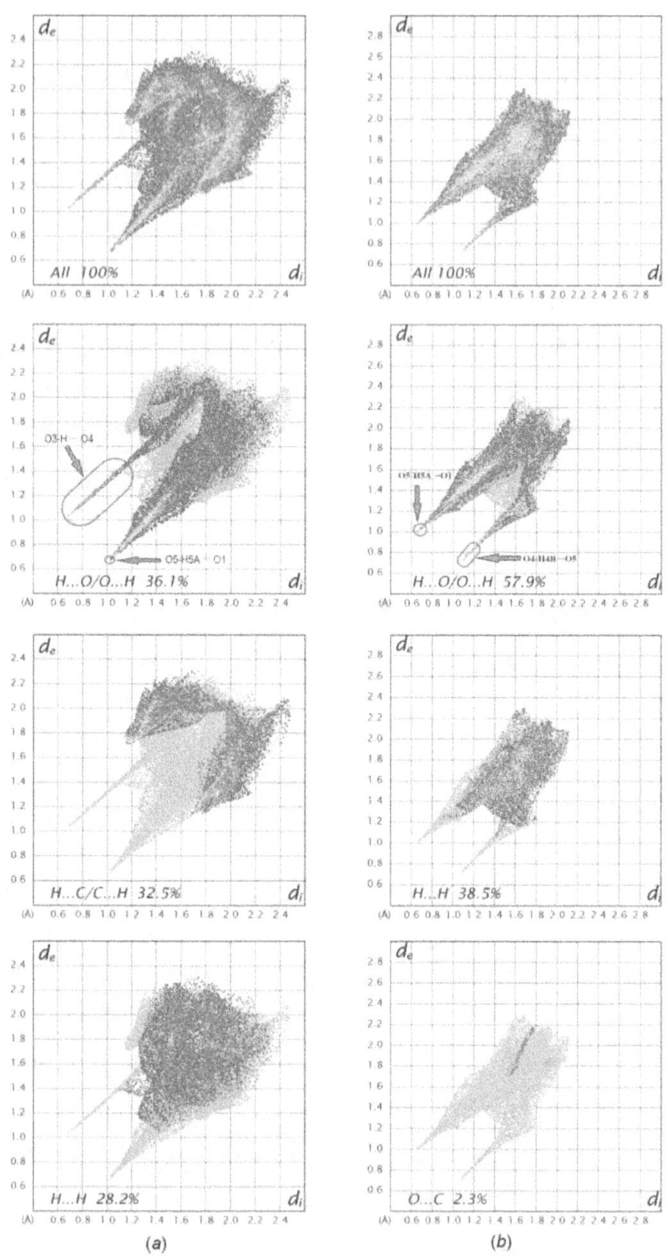

Figure 14. Two-dimensional fingerprint plots for (*a*) the PHB anion and (*b*) the [Ni(H$_2$O)$_6$]$^{2+}$ cation.

trans to each other) are engaged in three contacts, while the others exhibit only two contacts. The d_{norm} surfaces of the title compound include hydrogen bonding with the solvent water molecules, suggesting an increased stability of the hydrated form. The complete Hirshfeld surface analysis of the crystal structure shows that the major contribution to the inter- molecular interactions corresponds to strong H O/O H contacts. Fingerprint plots demonstrate that their contribu- tions are 36.1% for PHB and 57.9% for $[Ni(H_2O)_6]^{2+}$ (**Fig. 14**). Such a high percentage for the latter is unusual, but not unexpected considering that the complex ion contains six water molecules coordinated to the nickel center. Next in overall significance are the H H contacts, which contribute 28.2% and 38.5%, respectively, for the anionic and cationic fragments. However, in case of PHB, the contribution of H H contacts is smaller than the H C/C H contribution (32.5%) whereas the latter interaction is entirely insignificant in the cationic component. The percentage contribution of further weak interactions such as O C and C C is negligible.

4.2 HIRSHFELD SURFACE ANALYSIS OF [Cu(PHBA)₂(MEA)₂]

Any type of the intermolecular interactions can be described in detail by Hirshfeld surface analysis of the crystal structure. Molecular Hirshfeld surface over d_{norm} generated using a standard surface resolution and decomposed fingerprint plots of MC are presented in **Fig. 15**.

The d_{norm} surface was mapped on over the range of -0.7033 a.u.–1.3192 a.u, The d_{norm} mapping indicates N-H $\cdots$ O (for N1- H1A $\cdots$ O3 and N1-H1B $\cdots$ O2) and O-H $\cdots$ O (for

O3-H3$\cdots$O4) intermolecular hydrogen bonding interactions as bright red spots.

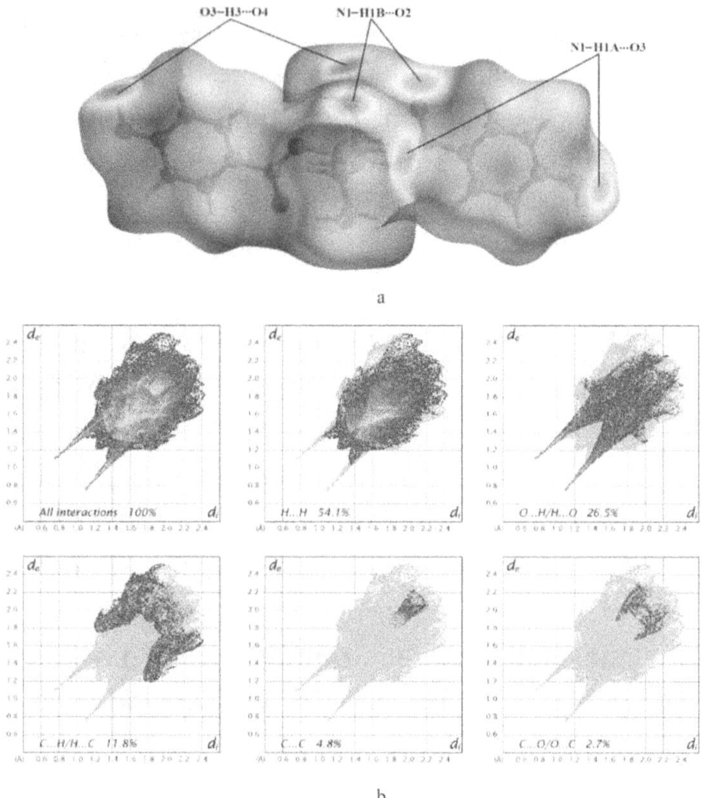

Fig. 15. Hirshfeld surface mapped for d_{norm} (a) and relative contributions of the various intermolecular contacts (b).

In the crystal packing of the compound, the most abundant intermolecular interactions observed globally were H$\cdots$H and O$\cdots$H/H$\cdots$O interactions. These interactions accounted for approximately 54.1% of the total surface area. The predominant H$\cdots$H contacts were formed between the hydrogens of methylene groups and benzene rings. This indicates that van der Waals forces play a significant role in stabilizing the crystal structure and influencing the packing arrangement of the molecules. The prevalence of these interactions highlights their importance in the overall structure and properties of the compound. The contribution of the $O \cdots H / H \cdots O$ interactions (mainly H-bonds) is 26.5% and they are indicated in finger print plot as sharp wings. Other interactions, *i.e.* $C \cdots H / H \cdots C$ (11.8%), $C \cdots C$ (4.8%) and $C \cdots O / O \cdots C$ (2.7%) contribute less to the Hirshfeld surface.

The C⋯H/H⋯C contacts show characteristic "wings" in the fingerprint plots which are identified as a result of weak C-H⋯π interactions. In order to characterize intermolecular contacts in more details the shape index and curvature are plotted in the range of -0.996–0.994 a.u. and -3.667 -0.224 a.u. (**Fig. 16 a** and **b**). The red triangles on the shape index represented by concave regions indicate π-stacking interactions whereas the blue triangles represented by convex regions indicate the ring atoms of the molecule inside the surface (**Fig. 16, a**). The red triangles on the shape index mapping refer to the C-H⋯π interaction with the contribution of 11.3% which is in accordance with the 2D fingerprint plot. C⋯C contacts attributed to π⋯π interactions between phenyl rings (C1-C6) were observed, with centroid-to-centroid distances of 3.701(2) Å. The corresponding fingerprint plots clearly depict a blue area on the diagonal at approximately 1.9 Å, which is characteristic of π⋯π interactions.

The curvedness indicates the electron density of surface curves around the molecular interactions (**Fig. 16, b**). The flat areas of the surface correspond to low value of curvedness, while sharp curvature area corresponds to high values of curvedness and usually tends to divide the surface into patches, indicating contacts between neighboring molecules. The large flat region delineated by a blue outline refers to π-π stacking interactions. Curvedness of the present compound indicates of π- π stacking interactions [83].

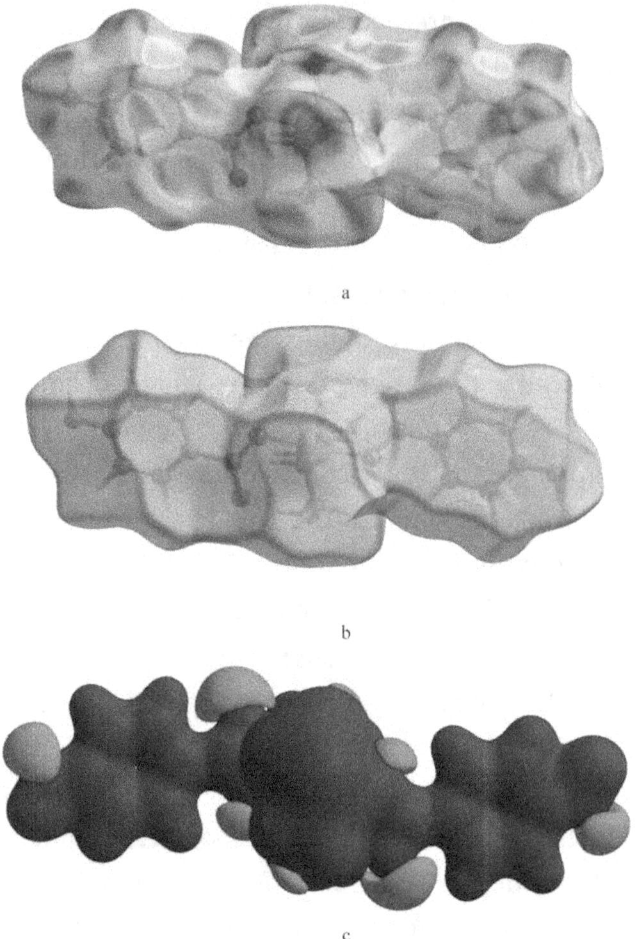

Fig. 16. Hirshfeld surfaces mapped for the shape index (a), curvedness (b) and view of the three-dimensional Hirshfeld surface plotted over electrostatic potential energy in the range-0.0500–0.0500 a.u. using the B3LYP/STO-3G basis set at the Hartree–Fock level of

theory. Hydrogen-bond donors and acceptors are shown as blue and red regions around the atoms corresponding to positive and negative potentials, respectively (c).

This finding is substantiated by the calculated electrostatic potential (**Fig. 16, c**) of the molecule that was used to generate the Hirshfeld surface. The negative potential (acceptor) is indicated as a red surface around the eight atoms (O1-O4 and O1_a-O4_a) and the blue surface area, indicating the positive potential (donor), is mapped in the proximity of the hydrogen atoms.

4.3 HIRSHFELD SURFACE ANALYSIS OF BINUCLEAR METAL COMPLEXES

Hirshfeld surface analysis is a very useful tool for visualizing interactions in molecular crystals. Using crystal structures of three complexes the intermolecular interactions are determined by Hirshfeld surface analysis. Molecular Hirshfeld surface over dnorm (normalize contact distance) generated using a standard surface resolution and decomposed fingerprint plots for $[Cu_2(o\text{-}HBA)_4(DMFA)_2]$ (complex-1), $[Cu_2(o\text{-}HBA)_4(H_2O)_2]DMFA$ (complex-2) and $\{[Cu_2(o\text{-}HBA)_2(p\text{-}HBA)_2(H_2O)_2]EtOH\}\{[Cu_2(o\text{-}HBA)_4(H_2O)_2]\}$ (complex-3) are presented in **Fig. 17-19** respectively [84–86]. Hirshfeld surface analysis for complex 1 is mapped over dnorm (normalize contact distance) in the color range from -0.703 (red) to1.319 (blue) a.u (**Fig.17**), which indicates the different close contacts with neighboring species of the complex molecule 1. The d_{norm} image is made transparent for visualizing the molecular structure of complex 1, where dnorm shows the different region of different color scheme, i.e., red, white and blue. Deep red spots are indicating the strong close interaction with the neighboring species that revealed classical O–H the crystal of complex 1. The proportion of O-H/H-O interactions comprising 12.1 and 10.7% of the Hirshfeld surface for complex- 1. For the entire complex-1 molecule, the broad region bearing short and narrow spikes at the middle of plot is reflected as H···H interaction in comprising 43% of the total Hirshfeld surfaces and this is the highest percentage calculated for H-H bond for the complex-1 and this is the usual phenomenon of Cu(II) complex of o/*p*-hydroxybenzoic acid [Fig. 12]. Other bonds of interaction comprising percentage are included in the **Fig.17.**

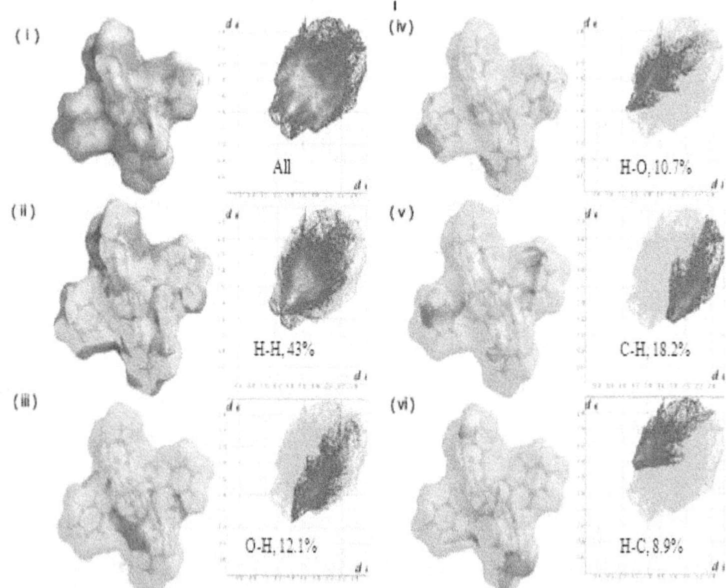

Fig 17. Hirshfeld surface mapped for d$_{norm}$, and 2D fingerprint plots of complex 1 and Relative contributions of the various intermolecular contacts of complex-1 [(i) All (ii)H-H,43% (iii)O-H,12.1% (iv)H-O,10.7 % (v)C-H, 18.2% (iv)H-C, 8.9%]

So, according to intermolecular interactions for complex-1, H···H, and C–H···π interaction with neighboring species are well dominated and are in complement to the Hirshfeld surfaces in the 2D fingerprint plots for complex- 1 (**Fig.17**).

In the complex-2, the interaction of H-H, O-H, H-O, C-H, and H-C in the surface of Hirshfeld were also calculated to be 38.5%, 13.9%, 11.0%, 18.2% and 8.9% respectively and usually highest percentage is observed for H-H bond **(Fig.18)**. From the Interaction intensity map, it is clearly observed the heavier the color of the red region, the stronger the O–H interactions. According to **Fig. 18** (III and IV) the O···H and H···O interactions have appeared symmetrical long sharp spikes in the 2D fingerprint plots. At the left of the plot, there are characteristic "wings" which are appeared as a result of C–H···π interactions like complex-1.

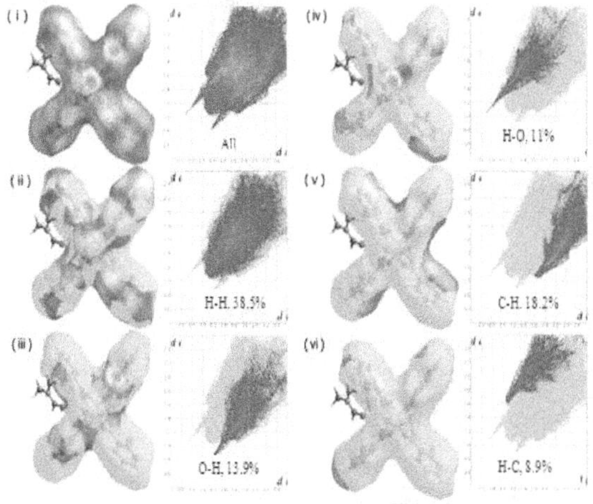

Fig 18. Hirshfeld surface mapped for d_{norm}, Ccomplex -2 and Relative contributions of the various intermolecular contacts of complex-2[(i) All (ii)H-H,38.5% (iii)O-H, 13.9 % (iv)H-O,11.0 % (v)C-H, 18.2% (iv)H-C, 8.9%]

In the complex-3, the percentage contribution to the Hirshfeld surface area by close contacts with H atoms inside the surface and H atoms outside was 40.5% and other interactions are also appeared in **Fig.19**. According to percentage of interaction, the H···H contacts was predominated of total Hirshfeld surface area of the molecules which was almost like complex-1 and 2. The O-H and H-O intermolecular interactions appeared as distinct spikes in the 2D fingerprint plot.

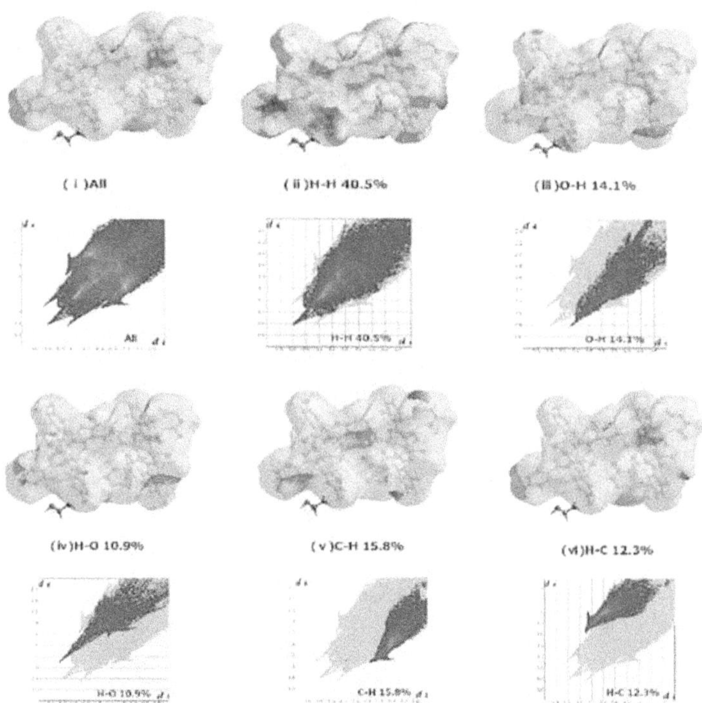

Fig. 19. Hirshfeld surface mapped for d_{norm}, complex -3 and relative contributions of the various intermolecular contacts of complex-3

PART 5. BIOLOGICAL ACTIVITIES OF SA/PHBA METAL COMPLEXES: AN OVERVIEW

5.1 INTRODUCTION

Nowadays due to the global climate changes and ecological problems a productivity of the agricultural plants is reducing. Therefore, it is of a crucial interest to suggest effective chemicals for accelerating a growth of plants and increasing their productivity. Moreover, it is important to control plant diseases caused by harmful microorganisms. In other words, suggested compounds have simultaneously to demonstrate plant growth and antimicrobial properties. Simple derivatives of benzoic acid may be used as such compounds.

Salicylic acid, $C_6H_4(OH)COOH$, where the OH group is *ortho*-position to the carboxyl group is a key ingredient in many skincare products for the treatment of seborrhoeic dermatitis, acne, psoriasis, calluses, corns, keratosis pilaris, acanthosis nigricans, ichthyosis and warts. Transition metal ions form many biologically important complexes with salicylates. They are widely used in medicine and have multiple effects on the metabolic processes [87]. The radioprotective effects of dihydrate diaqua- bis(salicylato)copper(II) complex (CuS), diaquabis(salicylato)zinc(II) complex (ZnS) and their mixture in molar ration 1:1 (CuS + ZnS) were assayed in mice [88].

Indeed, 4- or *para*-hydroxybenzoic acid (PHBA) which is a natural compound found in carrots, oil palm, grapes and etc. demonstrates a wide spectrum of the biological action such as antibacterial, antifungal, antialgal, antiviral and other activities [89–91] Unfortunately, a pronounced plant growth activity is not revealed for this derivative of benzoic acid.

Meanwhile, monoethanolamine (MEA) which can be found in a number of food items such as daikon radish, caraway, muscadine grape and etc. has noticeable plant growth [92] and antimicrobial [93] activity. MEA facilitates the survival of plants under abiotic and biotic stresses such as salinization [94] and growth stimulating effect is retained or even enhanced in its derivatives [95].

To enhance the growth stimulating action of p-hydroxybenzoic acid (PHBA), M. Crisan and colleagues have prepared a salt by combining this derivative of benzoic acid with monoethanolamine (MEA). The bioactivity tests conducted on the salt have demonstrated a significant improvement in its growth stimulating action. The enhanced performance of the salt suggests its potential application as a more effective growth stimulant. [96, 97] Our experiments evaluating the plant growth and antimicrobial properties of the salt formed between 4-nitrobenzoic acid and monoethanolamine (MEA) have yielded promising results. Despite the absence of significant fungicidal action from the acid component, we have observed enhancements in both plant growth and antimicrobial activities. This suggests that the salt formulation has the potential to be a valuable agent for promoting plant growth and exerting antimicrobial effects, even without the inherent fungicidal properties of the acid component. [98].

The other possible way of the biological activity enhancement is a preparation of metal complexes on the basis of benzoic acid derivative and MEA. In crystalline metal complexes hydrogen bonding and other weak interactions play a crucial role in self-assembly and recognition of the solid-state supramolecular structures [99–104] and obtained compound may demonstrate improved bioactivity. In this work we have synthesized mixed-ligand complex of copper with the PHBA and MEA (further MC), determined its molecular and crystal structure and growth activity on example of cotton plant.

In the below chapter practical experiments done by our "Supramolecular chemistry" group within grant project № F3-20200929348 will be given.

5.2 INFLUENCE OF MC TO COTTON PLANT GROWTH AND ENHANCEMENT OF PRODUCTIVITY (in the case of $[Cu(PHBA)_2(MEA)_2]$)

The ultimate goal of all physiological and agronomic researches is knowledge gaining about the most complicated mechanisms, laws of growth and development of plants, so that, on the basis of this knowledge to be able to create the most favorable conditions for the growth, development and production process of plants. Synthetic growth regulators allow to modify growth and development in the desired direction and the desired degree, exerting their action through endogenous levels of natural hormones. The use of these compounds in small doses leads to a significant production effect without changing the basic techniques of agricultural technology. The data of laboratory experiments obtained in order to determine the concentration of $[Cu(PHBA)_2(MEA)_2]$ (MC) (PHBA = p-hydroxybenzoic acid, MEA=monoethanolamine) metal complex (MC), at which the best germination of cotton seeds and optimal plant growth take place, are presented in **Table 14**.

Based on the data provided, the optimal concentration of the $[Cu(PHBA)_2(MEA)_2]$ (MC) complex is determined to be 0.001%. This concentration shows improved germination of cotton seeds compared to the control in plate experiments, as illustrated in Figure 20. These findings from the laboratory experiment confidently indicate that the studied compound possesses stimulating properties for the germination and development of cotton seeds.

As the next step, the influence of the prepared sample at the selected concentration was investigated on the growth and development of cotton plants in vegetation experiments. The yield of cotton, like other agricultural crops, is closely linked to the growth dynamics, formation of sympodial branches, and increase in leaf area, as these factors create conditions for the transition from vegetative growth to reproductive development. During the study, observations were made on the growth and development of plants throughout different vegetation phases. Parameters such as plant height and the number of branches were measured to analyze the impact of the growth regulator on these indicators. The results obtained are presented in Table 14.

Phenological observations indicate that there is a constant development of plants at the beginning of the vegetation period with the appearance of seedlings of cotton plants.

Table 14. Influence of MC on the dynamics of phenological indicators and cotton yield.

№	Sample		Indicator	1 Control	2 MC
	3–5 leaves	28.05.21	Height of the main stem, cm	16.25	19.75
	Budding	30.06.21	Height of the main stem, cm	78.5	91.25
			Number of sympodias, pcs/plant	13.00	16.75
	Fruiting	28.07.21	Number of buds,pcs/plant	17.75	20.75
			Height of the main stem, cm	100.75	103.25
			Number of sympodias, pcs/plant	13.55	17.75
	Ripening	08.09.2021	Number of buds,pcs/plant	18.0	24.75
			Number of bolls,pcs/plant	20	25.5
			Crop,gr/ vessel	70.0	94.35
			Weight of boll,gram	3.5	3.7
			Yield of crop,%	100	**134.8**

In contrast, a friendly and early appearance of seedlings was observed, both in the control variant (soaking the seeds with water, sowing, applying NPK) and in the tested variant (soaking seeds with the sample, sowing, applying NPK, spraying the plant leaves with MC).

The tallest and strongest seedlings were found in the variant using the preparation (3.5 cm higher than control). Due to weather conditions, the budding stage was characterized by rapid plant growth - the height of plants in the experimental variant reached 91.25 cm, 12.75 cm more than in control. The number of sympodial branches formed in the variant with the use of the preparation was also higher than in the variant with control plants by 3.75 branches, amounting to 16.75 and 13.0, respectively. A similar picture is observed with the formation of buds (**Table 14**).

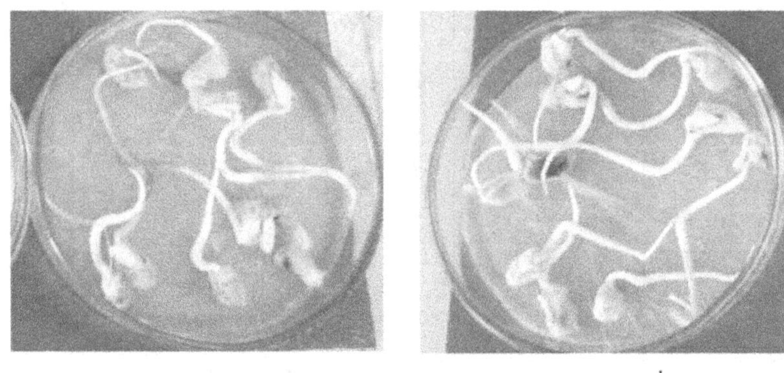

Fig. 20. The germination of cotton seeds in laboratory experiments: control (a) and treatment with MC (b).

In the next phase of development (fruiting at flowering phase), the following dynamics of plant growth is traced: in the experi-mental version, there is a slight slowdown in the growth of the main stem of plants, and in the control version, on the contrary, ac- celeration is observed. The formation of sympodial branches remains high in the experimental version with four more pieces. It is evident that MC had the most significant effect on the length of the main stem in the early phases of development. In the ripening phase in the test variant, the development of plants is clearly in the direction of the accumulation of reproductive organs (fruit elements), the number of bolls on 08.09.21 averaged 25.5 pcs/plant, while in the control variant this indicator reached 20 pcs.

The yield of raw cotton is the main criterion determining the effectiveness of the preparation. The results of field experiments attest that the yield of raw cotton in the test group exceeded the yield obtained in the variant of control plants (NPK) by 24.25 gr/vessel or almost 35% (**Table 14** and **Fig. 21, a**).

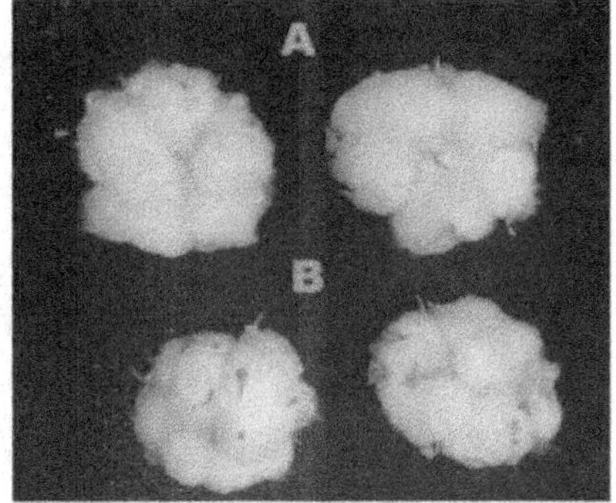

Fig. 21. Impact of the preparation on an enhancement of cotton productivity (left) and size and weight of cotton bolls
(right): (B) control; (A) plant treated by MC.

The stimulating effect of MC on the earlier opening of cotton bolls (3–5 days) also should be noted. It was recorded that in terms of weight, the weight of the bolls from the control variant (NPK) and the experimental group was practically at the same level - 3.5 and 3.7 gr/boll, respectively (**Fig. 21, b**).

Based on the abundance of determined indicators, it can be confidently concluded that the tested preparation exhibits excellent plant growth stimulation properties. Throughout the research period, MC at the determined concentration exerts a maximum stimulating effect and significantly impacts the root system of plants. Notably, it demonstrates a strong influence on the plant stem, leading to the development of additional sympodial branches.

The observed effects suggest that MC stimulates plant growth and development not only through auxin-like mechanisms but also exhibits gibberellin-like actions. These actions are believed to influence various physiological processes, including flowering, photosynthesis, respiration, water metabolism, nitrogen and carbohydrate metabolism, as well as nucleic acid metabolism.

Overall, the findings indicate that MC has the potential to serve as a multi-faceted growth regulator, influencing multiple aspects of plant physiology and promoting overall plant growth and development.

5.3 VALIDATING PRACTICAL EXPERIMENTS USING THE MOLECULAR DOCKING METHOD

In the realm of pharmaceutical research and drug development, the identification of potential drug candidates is a complex and time-consuming process. Traditional methods often fall short in predicting the binding affinity and interactions between small molecules and target proteins. However, with the advent of computational approaches, molecular docking has emerged as a valuable technique for investigating these interactions. Molecular docking enables scientists to simulate and evaluate the binding between a ligand (small molecule) and a receptor (target protein), providing crucial insights into drug design and discovery.

Applications of Molecular Docking:

a) Drug Discovery: Molecular docking plays a pivotal role in virtual screening, enabling scientists to explore vast libraries of small molecules and predict their binding affinities to specific target proteins. This aids in the identification of potential drug candidates with high binding affinity and favorable interactions, significantly accelerating the drug discovery process [105–116].

b) Protein Structure Prediction: Molecular docking can also be utilized to predict the binding modes between proteins, providing valuable insights into protein-protein interactions and aiding in the determination of protein structures [114, 117–126].

c) Understanding Protein-Ligand Interactions: By analyzing the docking results, researchers can gain a deeper understanding of the key interactions between a ligand and its target receptor. This knowledge aids in rational drug design, where modifications can be made to enhance the binding affinity and selectivity of potential drug candidates [127–139].

The molecular docking method allows studying ligands that bind to plant proteins. Considering this, in order to compare the [Cu(PHBA)₂(MEA)₂]-containing complex with positive results in plant growth with ligands and existing structurally similar natural auxins, Arabidopsis thaliana plant auxin receptor (AtTIR1) protein (PDB ID: 2P1P) bonding was made on the CB-DOCK server (China).

The Arabidopsis growth-responsive auxin acceptor transport inhibitor receptor 2p1p [140–144] (see **Figure 22**) was studied by individually exposing ligands and complexes. Protein-ligand interaction energies are presented in **Table 15**. Based on the data presented in the table, it was shown that the complexes can strongly bind to the protein in relation to the ligand molecules. This result is consistent with the results obtained during the agrochemical experiments presented in Chapter 5.2.

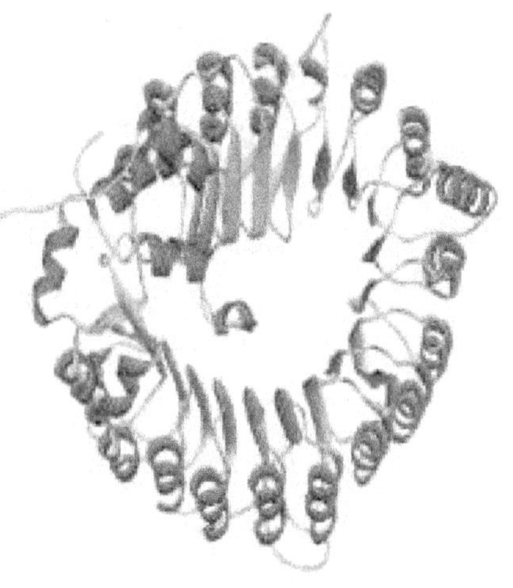

Figure 22. A 3D view of the Arabidopsis 2p1p protein

In calculations, the smallness of the energy in the impact processes is a factor that indicates the strength of the bond. A similar calculation was performed on PHBA and the [Cu(PHBA)₂(MEA)₂] complex synthesized with this ligand (see **Fig. 23**). The results of these calculations showed that the binding energy of PHBA to its own protein molecule alone (-5.3 kcal/mol) is greater than the binding energy of the complex compound synthesized through it (-7.1 kcal/mol) (see **Table 15**).

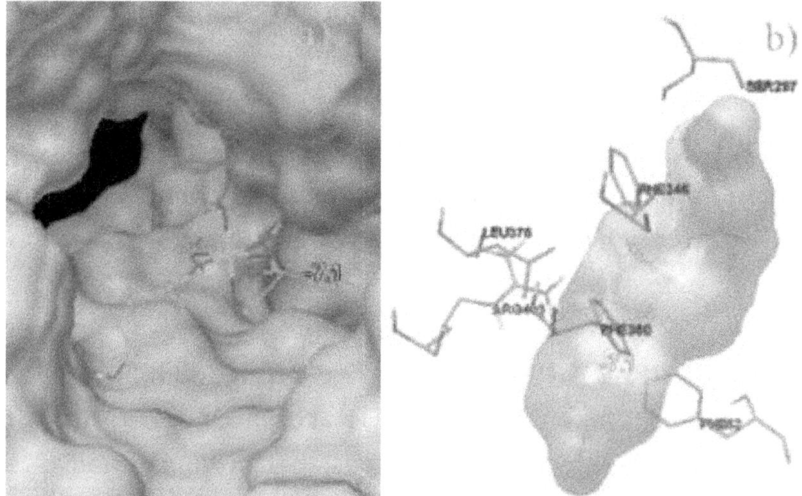

Fig. 23. a) position of [Cu(PHBA)₂(MEA)₂] in protein molecule;

b) Bonding of [Cu(PHBA)₂(MEA)₂] with amino acids

Table 15. Binding energies calculated by Autodock software

Checked compound	E$_{binding}$, (kcal/moll) (binding energy)	Center		
		x	y	z
SA	-5.5	16	-121	-21
MEA	-2.9	16	-121	-21
PHBA	-5.3	16	-121	-21
[Cu(PHBA)₂(MEA)₂]	-7.1	16	-121	-21
Phenylacetic acid	-5.6	16	-121	-21
2,4-dichlorophenoxyacetic acid	-6.7	16	-121	-21
Indole-3-butyric acid	-6.7	16	-121	-21

From the determined binding energies, it can be estimated that the binding energies of -5.6, -6.7, -6.7 kcal/mol corresponding to phenylacetic acid, 2,4-

dichlorophenoxyacetic acid, and Indole-3-butyric acid, which are considered natural auxins, are higher than the calculated binding energies for MEA, SA, PHBA and $Cu(SA)_2$.

$[Cu(PHBA)_2](MEA)_2]$ complex interacts with phenylalanine (PHE82, 280, 346), arginine (ARG403), serine (SER297) and leucine (LEU378) molecules. It was observed that these amino acid residues are connected with SER297 through hydrogen bonds between the hydroxyl group located in p-state.

The investigated complex compound $[Cu(PHBA)_2](MEA)_2]$ recorded stronger binding energies than auxins taken as a control. The compound $[Cu(PHBA)_2](MEA)_2]$, which showed the strongest performance, had a result 6% higher than the highest performing natural auxin.

5.4 ANTIMICROBIAL ACTIVITY OF $[Cu(PHBA)_2](MEA)_2]$

Previous laboratory studies to determine the properties of new complex compounds allowed to determine the effective concentration of the active substance, which positively affected the germination and germination power of cotton seeds, as well as the growth, development and yield of cotton in vegetation studies. Effective concentrations of 0.02% and 0.002% Cu+ PHBA +MEA preparate were also used in bacteriological tests to determine antimicrobial activity against microorganisms living in cotton seeds.

The experiment was also carried out at higher concentrations to determine the optimal concentration at which the effects of the drugs are most effective. Thus, concentrations of 2.0% and 4.0% of the active substance were chosen for Cu + PHBA + MEA. For comparative control, cotton seeds were soaked in distilled water and the drainage of this water was used for microbiological studies. The reason for choosing a high concentration of the studied preparations is based on the fact that the synthesized metal complexes have an antiseptic effect and that they are more effective in reducing bacterial contamination in cotton seeds.

Before microbiological studies, cotton seeds were soaked in water (control) and studied drugs in specified concentrations. Tests were carried out on cotton seeds of the Sultan elite variety. Seeds soaked in solutions were left at room temperature (24°C) for 24 hours.

To determine the antimicrobial activity of the studied drug after soaking the cotton seeds with a solution of the selected complex compound ($[Cu(PHBA)_2](MEA)_2]$), it was carried out three times in a universal cumulative medium - meat-peptone agar medium (MAP). After the completion of bacteriological inoculation, Petri dishes with seeds were placed in a thermostat and kept at the optimal temperature (26-27°C) for the growth of microorganism cells.

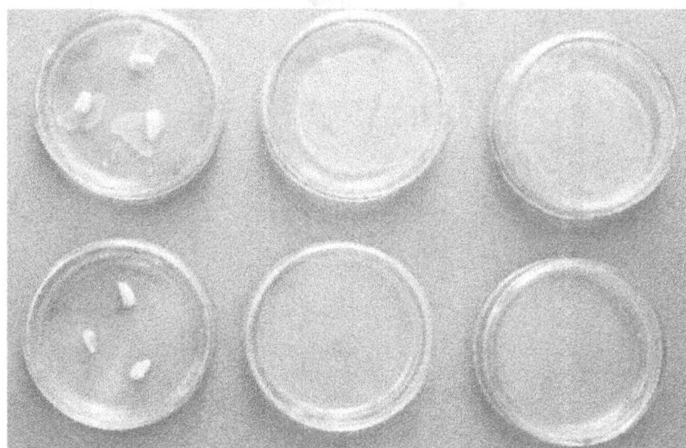

Figure 24. Antimicrobial effect of $[Cu(PHBA)_2(MEA)_2]$ at a concentration of 4% (bottom row) compared to control (top row).

When cotton seeds were pre-soaked in the studied preparations, the microorganism colonies were found in much lower numbers in the petri dish, which was particularly effective with a solution of [Cu(PHBA)$_2$](MEA)$_2$] at a concentration of 4.0% (see **Figure 24**).

Further studies to determine the antimicrobial activity of the studied substances showed that a decrease in the concentration of the active substance of the tested preparations led to a decrease in their ability to inhibit the growth of microorganisms. Thus, for [Cu(PHBA)$_2$(MEA)$_2$] at concentrations of 0.04% and 0.004%, the number of microorganisms is more than 51 million CFU/ml. Although these concentrations showed less antimicrobial activity compared to [Cu(PHBA)$_2$(MEA)$_2$] at 2.0% concentration, they were still more than 2-fold more effective than the control.

Figure 25. Effect of Cu+ PHBA +MEA at 0.004% concentration (right side), seed deterioration process compared to control (left side).

Cotton seeds were left in beaker, where they were treated with solutions of the studied substance to determine the time of decay of the seed. On the seventh day, the germination of cotton seeds was observed in the beaker, in which solutions with a concentration of [Cu(PHBA)$_2$(MEA)$_2$] – 0.004% were particularly effective (see **Figure 25**). During this period, the seeds left in water for the control died during the spoilage process.

5.5 CONCLUSION

According to the results of the research, it can be concluded that the [Cu(PHBA)$_2$(MEA)$_2$] complex solutions at concentrations of 4% and 2% are effective in having the most inhibitory effect on growth and development. Decreasing the concentration of [Cu(PHBA)$_2$(MEA)$_2$] (0.04%), although it caused an increase in microorganisms, was still more effective than the control option. It was noted as an acceptable option for pre-sowing seed treatment in crop treatment.

Application of 0.0002% solution of the complex for foliar feeding has been tested in practice to increase productivity, it is recommended for use in cotton cultivation.

The use of dual-effect drugs in increasing the phenological features of the cotton plant and the final yield allows to collect the forecasted grain in the agricultural sector in higher than expected quantities. Metal complexes used in very small amounts bring the plant into a state of microstress and increase its resistance to external and internal influences and cause faster development at all stages of growth.

REFERENCES

1. Zeb, A. (2021). A comprehensive review on different classes of polyphenolic compounds present in edible oils. *Food Research International*. Elsevier Ltd. https://doi.org/10.1016/j.foodres.2021.110312

2. Gao, Z., Ge, C., Baker, R. C., Tikekar, R. V, & Buchanan, R. L. (2022). Enhancement of Thermal Inactivation of Cronobacter sakazakii in Apple Juice at 58°C by Inclusion of Butyl Para-Hydroxybenzoate and Malic Acid. *Journal of Food Protection*, *85*(11), 1515–1521. https://doi.org/https://doi.org/10.4315/JFP-22-039

3. Gao, Z., Ge, C., Baker, R. C., Tikekar, R. V, & Buchanan, R. L. (2022). Evaluation of Potential for Butyl and Heptyl Para-Hydroxybenzoate Enhancement of Thermal Inactivation of Cronobacter sakazakii during Rehydration of Powdered Infant Formula and Nonfat Dry Milk. *Journal of Food Protection*, *85*(8), 1133–1141. https://doi.org/https://doi.org/10.4315/JFP-22-044

4. Gao, Z., Ding, Q., Ge, C., Baker, R. C., Tikekar, R. V, & Buchanan, R. L. (2021). Synergistic Effects of Butyl Para-Hydroxybenzoate and Mild Heating on Foodborne Pathogenic Bacteria. *Journal of Food Protection*, *84*(4), 545–552. https://doi.org/https://doi.org/10.4315/JFP-20-175

5. Ailincai, D., Marin, L., Shova, S., & Tuchilus, C. (2016). Benzoate liquid crystals with direct isotropic–smectic transition and antipathogenic activity. *Comptes Rendus Chimie*, *19*(5), 556–565.

https://doi.org/https://doi.org/10.1016/j.crci.2016.01.008

6. K., S. F., M., S., & K., A. (2018). Structural investigation and molecular docking studies of 2-amino-1H-benzimidazolium 2-hydroxybenzoate and 2-amino-1H-benzimidazolium pyridine 2-carboxylate single crystal. *Materials Science and Engineering: C, 91*, 103–114. https://doi.org/https://doi.org/10.1016/j.msec.2018.05.035

7. Zhang, M., Han, W., Hu, X., Li, D., Ma, X., Liu, H., … Liu, S. (2020). Pentaerythritol p-hydroxybenzoate ester-based zinc metal alkoxides as multifunctional antimicrobial thermal stabilizer for PVC. *Polymer Degradation and Stability, 181*, 109340. https://doi.org/https://doi.org/10.1016/j.polymdegradstab.2020.109340

8. Saiadali Fathima, K., Sathiyendran, M., & Anitha, K. (2019). Structure elucidation, biological evaluation and molecular docking studies on a new organic salt: 2-Aminobenzimidazolium 5-nitro-2-hydroxybenzoate. *Journal of Molecular Structure, 1177*, 457–468. https://doi.org/https://doi.org/10.1016/j.molstruc.2018.09.059

9. Zhang, J., Ma, J., Jiao, G., Liu, K., Cui, R., Zhai, S., & Sun, R. (2023). Methyl 4-hydroxybenzoate nanospheres anchored on poly(amidoxime)/polyvinyl alcohol hydrogel network with excellent antibacterial activity for efficient uranium extraction from seawater. *Desalination, 548*, 116243. https://doi.org/https://doi.org/10.1016/j.desal.2022.116243

10.   Ruzmetov, A. K., Ibragimov, A. B., Myachina, O. V., Kim, R. N., Mamasalieva, L. E., Ashurov, J. M., & Ibragimov, B. T. (2022). Synthesis, crystal structure, Hirshfeld surface analysis and bioactivity of the Cu mixed-ligand complex with 4-hydroxybenzoic acid and monoehtanolamine. *Chemical Data Collections*, *38*(January), 100845, 1–11. https://doi.org/10.1016/j.cdc.2022.100845

11.   Shen, Y., Sun, F., Zhang, L., Cheng, Y., Zhu, H., Wang, S.-P., … Lin, H.-W. (2020). Biosynthesis of depsipeptides with a 3-hydroxybenzoate moiety and selective anticancer activities involves a chorismatase. *Journal of Biological Chemistry*, *295*(16), 5509–5518. https://doi.org/https://doi.org/10.1074/jbc.RA119.010922

12.   Salehi, B., Vlaisavljevic, S., Adetunji, C. O., Adetunji, J. B., Kregiel, D., Antolak, H., … Contreras, M. del M. (2019). Plants of the genus Vitis: Phenolic compounds, anticancer properties and clinical relevance. *Trends in Food Science & Technology*, *91*, 362–379. https://doi.org/https://doi.org/10.1016/j.tifs.2019.07.042

13.   Slaihim, M. M., Al-Suede, F. S. R., Khairuddean, M., Khadeer Ahamed, M. B., & Shah Abdul Majid, A. M. (2019). Synthesis, characterisation of new derivatives with mono ring system of 1,2,4-triazole scaffold and their anticancer activities. *Journal of Molecular Structure*, *1196*, 78–87. https://doi.org/https://doi.org/10.1016/j.molstruc.2019.06.066

14.   Durai, P., Chinnasamy, A., Gajendran, B., Ramar, M.,

Pappu, S., Kasivelu, G., & Thirunavukkarasu, A. (2014). Synthesis and characterization of silver nanoparticles using crystal compound of sodium para-hydroxybenzoate tetrahydrate isolated from Vitex negundo. L leaves and its apoptotic effect on human colon cancer cell lines. *European Journal of Medicinal Chemistry, 84*, 90–99. https://doi.org/https://doi.org/10.1016/j.ejmech.2014.07.012

15.   Mortikov, V. Y., Litvinov, V. P., Shestopalov, A. M., Sharanin, Y. A., Apenova, E. é., Galegov, G. A., … Abdullaev, F. I. (1991). Synthesis and antiviral activity of 3-cyano-2(1H)pyridine selenones. *Pharmaceutical Chemistry Journal, 25*(5), 312–317. https://doi.org/10.1007/BF00772121

16.   Hamidah, I., Sriyono, & Hudha, M. N. (2020). A bibliometric analysis of COVID-19 research using vosviewer. *Indonesian Journal of Science and Technology, 5*(2), 209–216. https://doi.org/10.17509/ijost.v5i2.24522

17.   van Eck, N. J., & Waltman, L. (2010). Software survey: VOSviewer, a computer program for bibliometric mapping. *Scientometrics, 84*(2), 523–538. https://doi.org/10.1007/s11192-009-0146-3

18.   Prahani, B. K., Alfin, J., Fuad, A. Z., Saphira, H. V., Hariyono, E., & Suprapto, N. (2022). Learning Management System (LMS) Research During 1991–2021: How Technology Affects Education. *International Journal of Emerging Technologies in Learning, 17*(17), 28–49. https://doi.org/10.3991/ijet.v17i17.30763

19.   Lewandowski, W., Kalinowska, M., & Lewandowska, H.

(2005). The influence of metals on the electronic system of biologically important ligands. Spectroscopic study of benzoates, salicylates, nicotinates and isoorotates. Review. *Journal of Inorganic Biochemistry, 99*(7), 1407–1423. https://doi.org/10.1016/j.jinorgbio.2005.04.010

20.   Egorov, V. M., Djigailo, D. I., Momotenko, D. S., Chernyshov, D. V., Torocheshnikova, I. I., Smirnova, S. V., & Pletnev, I. V. (2010). Task-specific ionic liquid trioctylmethylammonium salicylate as extraction solvent for transition metal ions. *Talanta, 80*(3), 1177–1182. https://doi.org/10.1016/j.talanta.2009.09.003

21.   Sayyari, M., Babalar, M., Kalantari, S., Martínez-Romero, D., Guillén, F., Serrano, M., & Valero, D. (2011). Vapour treatments with methyl salicylate or methyl jasmonate alleviated chilling injury and enhanced antioxidant potential during postharvest storage of pomegranates. *Food Chemistry, 124*(3), 964–970. https://doi.org/10.1016/j.foodchem.2010.07.036

22.   Liu, T., Song, T., Zhang, X., Yuan, H., Su, L., Li, W., … Dou, D. (2014). Unconventionally secreted effectors of two filamentous pathogens target plant salicylate biosynthesis. *Nature Communications, 5*(1), 4686. https://doi.org/10.1038/ncomms5686

23.   Essandoh, M., Kunwar, B., Pittman, C. U., Mohan, D., & Mlsna, T. (2015). Sorptive removal of salicylic acid and ibuprofen from aqueous solutions using pine wood fast pyrolysis biochar. *Chemical Engineering Journal, 265*, 219–

227. https://doi.org/10.1016/j.cej.2014.12.006

24.   Ranjit, K. ., Willner, I., Bossmann, S. ., & Braun, A. . (2001). Lanthanide Oxide Doped Titanium Dioxide Photocatalysts: Effective Photocatalysts for the Enhanced Degradation of Salicylic Acid and t-Cinnamic Acid. *Journal of Catalysis*, *204*(2), 305–313. https://doi.org/10.1006/jcat.2001.3388

25.   Shen, R., & Porco, J. A. (2000). Synthesis of Enamides Related to the Salicylate Antitumor Macrolides Using Copper-Mediated Vinylic Substitution. *Organic Letters*, *2*(9), 1333–1336. https://doi.org/10.1021/ol005800t

26.   Tong, M.-L., Chen, H.-J., & Chen, X.-M. (2000). Molecular Ladders with Multiple Interpenetration of the Lateral Arms into the Squares of Adjacent Ladders Observed for [M2(4,4′-bpy)3(H2O)2(phba)2](NO3)2·4H2O (M = Cu2+ or Co2+; 4,4′-bpy = 4,4′-Bipyridine; phba = 4-Hydroxybenzoate). *Inorganic Chemistry*, *39*(10), 2235–2238. https://doi.org/10.1021/ic991312+

27.   Li, N., Han, X., Feng, D., Yuan, D., & Huang, L.-J. (2019). Signaling Crosstalk between Salicylic Acid and Ethylene/Jasmonate in Plant Defense: Do We Understand What They Are Whispering? *International Journal of Molecular Sciences*, *20*(3), 671. https://doi.org/10.3390/ijms20030671

28.   Scarabelli, L., Grzelczak, M., & Liz-Marzán, L. M. (2013). Tuning Gold Nanorod Synthesis through Prereduction with Salicylic Acid. *Chemistry of Materials*, *25*(21), 4232–4238.

https://doi.org/10.1021/cm402177b

29.   Liu, L., Sonbol, F.-M., Huot, B., Gu, Y., Withers, J., Mwimba, M., … Dong, X. (2016). Salicylic acid receptors activate jasmonic acid signalling through a non-canonical pathway to promote effector-triggered immunity. *Nature Communications*, *7*(1), 13099. https://doi.org/10.1038/ncomms13099

30.   Chan, Z., Qin, G., Xu, X., Li, B., & Tian, S. (2007). Proteome Approach To Characterize Proteins Induced by Antagonist Yeast and Salicylic Acid in Peach Fruit. *Journal of Proteome Research*, *6*(5), 1677–1688. https://doi.org/10.1021/pr060483r

31.   Shi, Y.-L., & Shi, M. (2007). The synthesis of chromenes, chromanes, coumarins and related heterocycles via tandem reactions of salicylic aldehydes or salicylic imines with α,β-unsaturated compounds. *Org. Biomol. Chem.*, *5*(10), 1499–1504. https://doi.org/10.1039/B618984A

32.   Colón, G., Hidalgo, M. ., & Navío, J. . (2001). Photocatalytic deactivation of commercial TiO2 samples during simultaneous photoreduction of Cr(VI) and photooxidation of salicylic acid. *Journal of Photochemistry and Photobiology A: Chemistry*, *138*(1), 79–85. https://doi.org/10.1016/S1010-6030(00)00372-5

33.   Bica, K., Rijksen, C., Nieuwenhuyzen, M., & Rogers, R. D. (2010). In search of pure liquid salt forms of aspirin: ionic liquid approaches with acetylsalicylic acid and salicylic acid. *Physical Chemistry Chemical Physics*, *12*(8), 2011.

https://doi.org/10.1039/b923855g

34.   BABALAR, M., ASGHARI, M., TALAEI, A., & KHOSROSHAHI, A. (2007). Effect of pre- and postharvest salicylic acid treatment on ethylene production, fungal decay and overall quality of Selva strawberry fruit. *Food Chemistry*, *105*(2), 449–453. https://doi.org/10.1016/j.foodchem.2007.03.021

35.   Yates, C. R., Benítez, D., Khan, S. I., Stoddart, J. F., Komine, Y., & Tanaka, K. (2010). Organic letters化合物データテンプレート, *1569*(I), 1. Opgehaal van https://pubs.acs.org/journal/orlef7

36.   Ioannidis, J. P. A. (2006). Concentration of the Most-Cited Papers in the Scientific Literature: Analysis of Journal Ecosystems. *PLoS ONE*, *1*(1), e5. https://doi.org/10.1371/journal.pone.0000005

37.   Jalal, S. K. (2019). Co-authorship and co-occurrences analysis using bibliometrix r-package: A case study of india and bangladesh. *Annals of Library and Information Studies*, *66*(2), 57–64.

38.   van der Horn, J., Souvignier, B., & Lutz, M. (2017). Crystallization, Structure Determination and Reticular Twinning in Iron(III) Salicylate: Fe[(HSal)(Sal)(H2O)2]. *Crystals*, *7*(12), 377. https://doi.org/10.3390/cryst7120377

39.   Сергиенко, В. С., Абраменко, В. Л., & Суражская, М. Д. (2020). Синтез, кристаллическая и молекулярная структура сольватированного комплекса [MoO · C .

Кристаллография, *65*(4), 593–596.
https://doi.org/10.31857/S0023476120030303

40.   Sheldrick, G. M., & IUCr. (2007). A short history of SHELX. *urn:issn:0108-7673*, *64*(1), 112–122.
https://doi.org/10.1107/S0108767307043930

41.   Ashurov, Z. M., Tashkhodjaev, B., Izotova, L. Y., Olimova, M. I., & Ibragimov, B. T. (2017). Crystal and molecular structure of β-(N-benzoxazoline-2-thione) propionic acid and its salts. *Journal of Structural Chemistry*, *58*(3), 544–549. https://doi.org/10.1134/S0022476617030167/METRICS

42.   Sheldrick, G. M., & IUCr. (2015). Crystal structure refinement with SHELXL. *urn:issn:2053-2296*, *71*(1), 3–8.
https://doi.org/10.1107/S2053229614024218

43.   Macrae, C. F., Bruno, I. J., Chisholm, J. A., Edgington, P. R., McCabe, P., Pidcock, E., … Wood, P. A. (2008). Mercury CSD 2.0 - New features for the visualization and investigation of crystal structures. *Journal of Applied Crystallography*, *41*(2), 466–470. https://doi.org/10.1107/S0021889807067908

44.   Ibragimov, A., Englert, U., … J. A.-J. of S., & 2018, undefined. (2018). Dimorphism of Hexaaquanickel(Ii) Bis(p-Nitrobenzoate) Dihydrate Salt: A New Triclinic Crystal Form. *Springer*, *59*(2), 411–414.
https://doi.org/10.1134/S0022476618020221

45.   Bednarchuk, T. J., Kinzhybalo, V., & Pietraszko, A. (2016). Synthesis, structure and characterization of five new organically templated metal sulfates with 2-aminopyridinium.

Acta Crystallographica Section C: Structural Chemistry, 72(5), 432–441. https://doi.org/10.1107/S2053229616006458

46.   Sheldrick, G. M., & IUCr. (2015). SHELXT – Integrated space-group and crystal-structure determination. *urn:issn:2053-2733, 71*(1), 3–8. https://doi.org/10.1107/S2053273314026370

47.   Grell, J., Bernstein, J., & Tinhofer, G. (1999). Graph-set analysis of hydrogen-bond patterns: some mathematical concepts. *urn:issn:0108-7681, 55*(6), 1030–1043. https://doi.org/10.1107/S0108768199007120

48.   Crisan, M., Bourosh, P., Chumakov, Y., Petric, M., & Ilia, G. (2013). Supramolecular assembly and Ab initio quantum chemical calculations of 2-hydroxyethylammonium salts of para-substituted benzoic acids. *Crystal Growth and Design, 13*(1), 143–154. https://doi.org/10.1021/CG301304Y/SUPPL_FILE/CG301304Y _SI_002.CIF

49.   Ibragimov, A. B., Ashurov, Z. M., & Zakirov, B. S. (2017). Molecular and crystal structure of a mixed-ligand cadmium complex with p-hydroxybenzoic acid and monoethanolamine. *Journal of Structural Chemistry, 58*(3), 588–590. https://doi.org/10.1134/S0022476617030209/METRICS

50.   Cheng, H. M., Gao, X. W., Zhang, K., Wang, X. R., Zhou, W., Li, S. J., … Yan, D. P. (2019). A novel antimicrobial composite: ZnAl-hydrotalcite with: P -hydroxybenzoic acid intercalation and its possible application as a food packaging

material. *New Journal of Chemistry*, *43*(48), 19408–19414. https://doi.org/10.1039/c9nj03943k

51.   Olczak-Kobza, M., & Mrozek, A. (2009). Zinc(II) and cadmium(II) complexes with o-hydroxybenzoic acid or o-aminobenzoic acid and 2-methylimidazole: IIIR spectra, X-ray diffraction studies and thermal analysis. *Journal of Thermal Analysis and Calorimetry*, *96*(2), 553–560. https://doi.org/10.1007/s10973-008-9048-5

52.   Ibragimov, A. B., Ashurov, Z. M., & Zakirov, B. S. (2017). Two supramolecular complexes based on 4-hydroxybenzoic acid and triethanolamine: Synthesis and structure. *Russian Journal of Inorganic Chemistry*, *62*(4), 439–445. https://doi.org/10.1134/S0036023617040064

53.   Ruzmetov, A. K., Ibragimov, A. B., Myachina, O. V., Kim, R. N., Mamasalieva, L. E., Ashurov, J. M., & Ibragimov, B. T. (2022). Synthesis, crystal structure, Hirshfeld surface analysis and bioactivity of the Cu mixed-ligand complex with 4-hydroxybenzoic acid and monoehtanolamine. *Chemical Data Collections*, *38*, 100845. https://doi.org/10.1016/J.CDC.2022.100845

54.   Harrison, W., Rettig, S., & Trotter, J. (1972). Crystal and molecular structure of tetra-μ-o-bromobenzoato-bis[aquo-copper(II)]. *Journal of the Chemical Society, Dalton Transactions*, (17), 1852–1856. https://doi.org/10.1039/DT9720001852

55.   Melník, M. (1982). Study of the relation between the structural data and magnetic interaction in oxo-bridged

binuclear copper(II) compounds. *Coordination Chemistry Reviews, 42*(2), 259–293. https://doi.org/10.1016/S0010-8545(00)80537-8

56.   Gorincioy, V. V., Simonov, Y. A., Shova, S. G., Shofranskii, V. N., & Turta, C. I. (2009). Crystal and molecular structures of binuclear complexes of salycilic acid {Cu-M} (M = Cu, Sr, Ba). *Journal of Structural Chemistry, 50*(6), 1143–1148. https://doi.org/10.1007/S10947-009-0167-Z/METRICS

57.   Layana, S. R., Saritha, S. R., Anitha, L., Sithambaresan, M., Sudarsanakumar, M. R., & Suma, S. (2018). Synthesis, spectral characterization and structural studies of a novel O, N, O donor semicarbazone and its binuclear copper complex with hydrogen bond stabilized lattice. *Journal of Molecular Structure, 1157*, 579–586. https://doi.org/10.1016/J.MOLSTRUC.2017.12.073

58.   Desiraju, G. R. (1996). The C-H⋯O Hydrogen Bond: Structural Implications and Supramolecular Design. *Accounts of Chemical Research, 29*(9), 441–449. https://doi.org/10.1021/AR950135N/ASSET/AR950135N.FP.P NG_V03

59.   Desiraju, G. R. (2002). Hydrogen Bridges in Crystal Engineering: Interactions without Borders. *Accounts of Chemical Research, 35*(7), 565–573. https://doi.org/10.1021/AR010054T

60.   Guo, C., Zhang, Q., Zhu, B., Zhang, Z., Bao, J., Ding, Q., … Mei, X. (2020). Pharmaceutical Cocrystals of Nicorandil with Enhanced Chemical Stability and Sustained Release. *Crystal*

Growth and Design, 20(10), 6995–7005.
https://doi.org/10.1021/acs.cgd.0c01043

61.   Rigaku, O. D. (2020). CrysAlis PRO. Rigaku Oxford
Diffraction, Rigaku Corporation. Oxford, England.

62.   Melník, M., Dunaj-Jurčo, M., & Handlovič, M. (1984).
Crystal structure, spectral and magnetic characterization of
bis(μ-benzoato-O, O') (dimethylsulphoxide)copper(II).
Inorganica Chimica Acta, 86(3), 185–190.
https://doi.org/10.1016/S0020-1693(00)83768-8

63.   Klinga, M., Sundberg, M. R., Melník, M., & Mroziński, J.
(1989). Spectroscopic, magnetic and structural
characterization of bis(dimethyl sulfoxide)tetrakis(μ-2-
nitrobenzoato-O,O')dicopper(II). *Inorganica Chimica Acta,
162*(1), 39–43. https://doi.org/10.1016/S0020-
1693(00)83117-5

64.   Kristiansson, O., & Tergenius, L. E. (2001). Structure
and host–guest properties of the nanoporous
diaquatetrakis(p-nitrobenzoato)dicopper(II) framework.
Journal of the Chemical Society, Dalton Transactions, (9),
1415–1420. https://doi.org/10.1039/B100734N

65.   Muhonen, H. (1986). Exchange Interaction through
Hydrogen-Bond Bridges and the Effect of a Single-Oxygen
Bridge. Crystal Structures and Magnetic Susceptibilities of Two
Binuclear Copper(II) Complexes of 2-Amino-2-methyl-l-
propanol. *Inorganic Chemistry, 25*(26), 4692–4698.
https://doi.org/10.1021/IC00246A021/SUPPL_FILE/IC00246A0
21_SI_001.PDF

66.   O'Connor, M., Kellett, A., McCann, M., Rosair, G., McNamara, M., Howe, O., … Devereux, M. (2012). Copper(II) complexes of salicylic acid combining superoxide dismutase mimetic properties with DNA binding and cleaving capabilities display promising chemotherapeutic potential with fast acting in vitro cytotoxicity against cisplatin sensitive and resista. *Journal of Medicinal Chemistry*, *55*(5), 1957–1968. https://doi.org/10.1021/jm201041d

67.   Karipides, A., & White, C. (1993). Structure of diaquatetrakis(μ-2,6-difluorobenzoato-κO:κO′)dicopper(II). *urn:issn:0108-2701*, *49*(11), 1920–1923. https://doi.org/10.1107/S0108270193004214

68.   Gorinchoi, V. V., Turte, K. I., Simonov, Y. A., Shova, S. G., Lipkovskii, Y., & Shofranskii, V. N. (2009). Heteronuclear {Fe-Ba, Fe-Sr} salicylate complexes. Synthesis, structure, and physicochemical properties. *Russian Journal of Coordination Chemistry/Koordinatsionnaya Khimiya*, *35*(4), 279–285. https://doi.org/10.1134/S1070328409040083

69.   You, Z. L., & Zhu, H. L. (2004). Syntheses, Crystal Structures, and Antibacterial Activities of Four Schiff Base Complexes of Copper and Zinc. *Zeitschrift für anorganische und allgemeine Chemie*, *630*(15), 2754–2760. https://doi.org/10.1002/ZAAC.200400270

70.   Stachová, P., Valigura, D., Koman, M., Melník, M., Korabik, M., Mroziński, J., & Glowiak, T. (2004). Crystal structure, magnetic and spectral properties of tetrakis(2-nitrobenzoato)di(aqua)dicopper(II) dihydrate. *Polyhedron*,

23(8), 1303–1308.
https://doi.org/10.1016/J.POLY.2003.12.025

71.   Wang, W. H., Liu, W. S., Wang, Y. W., Li, Y., Zheng, L. F., & Wang, D. Q. (2007). Self-assembly and cytotoxicity study of waterwheel-like dinuclear metal complexes: The first metal complexes appended with multiple free hydroxamic acid groups. *Journal of Inorganic Biochemistry*, *101*(2), 297–304. https://doi.org/10.1016/J.JINORGBIO.2006.10.002

72.   Fu, X. P., Shen, J. W., Chen, L., Zhong, D. X., Wang, Y. L., & Liu, Q. Y. (2023). Dicopper(II) paddle-wheel metal–organic frameworks for high propyne storage under ambient conditions. *Chemical Communications*, *59*(16), 2263–2266. https://doi.org/10.1039/D2CC06684J

73.   Golomb, M. J., Calbo, J., Bristow, J. K., & Walsh, A. (2020). Ligand engineering in Cu(II) paddle wheel metal–organic frameworks for enhanced semiconductivity. *Journal of Materials Chemistry A*, *8*(26), 13160–13165. https://doi.org/10.1039/D0TA04466K

74.   Totaro, G., Sisti, L., Celli, A., Askanian, H., Hennous, M., Verney, V., & Leroux, F. (2017). Chain extender effect of 3-(4-hydroxyphenyl)propionic acid/layered double hydroxide in PBS bionanocomposites. *European Polymer Journal*, *94*, 20–32. https://doi.org/10.1016/J.EURPOLYMJ.2017.06.031

75.   Del Arco, M., Cebadera, E., Gutiérrez, S., Martín, C., Montero, M. J., Rives, V., … Sevilla, M. A. (2004). Mg,Al layered double hydroxides with intercalated indomethacin: Synthesis, characterization, and pharmacological study.

Journal of Pharmaceutical Sciences, 93(6), 1649–1658. https://doi.org/10.1002/jps.20054

76.   Lin, C. J., Zheng, Y. Q., Zhang, D. Y., & Xu, W. (2014). Syntheses, crystal structures, and characterization of copper(II) carboxylate complexes incorporating o-hydroxybenzoic acid and p-hydroxybenzoic acid. *Russian Journal of Coordination Chemistry/Koordinatsionnaya Khimiya, 40*(12), 932–942. https://doi.org/10.1134/S1070328414120094/METRICS

77.   Rissanen, K., Valkonen, J., Kokkonen, P., Leskelä, M., & Niinistö, L. (1987). Structural and Thermal Studies on Salicylato Complexes of Divalent Manganese, Nickel, Copper and Zinc. *Acta Chemica Scandinavica, 41a*, 299–309. https://doi.org/10.3891/ACTA.CHEM.SCAND.41A-0299

78.   Timakova, E. V., Udalova, T. A., & Yukhin, Y. M. (2009). Precipitation of bismuth(III) salicylates from mineral acid solutions. *Russian Journal of Inorganic Chemistry, 54*(6), 873–880. https://doi.org/10.1134/S0036023609060096

79.   Abuhijleh, A. L., & Woods, C. (2001). Mononuclear copper (II) salicylate imidazole complexes derived from copper (II) aspirinate. Crystallographic determination of three copper geometries in a unit cell. *Inorganic Chemistry Communications, 4*(3), 119–123. https://doi.org/10.1016/S1387-7003(00)00221-5

80.   Al-Obaidi, O. H. S. (2012). Binuclear Cu(II) and Co(II) complexes of tridentate heterocyclic shiff base derived from salicylaldehyde with 4-aminoantipyrine. *Bioinorganic*

Chemistry and Applications, 2012.
https://doi.org/10.1155/2012/601879

81.   Harvey, P. D., & Gray, H. B. (1988). Low-Lying Singlet and Triplet Electronic Excited States of Binuclear (d10-d10) Palladium(0) and Platinum(0) Complexes. *Journal of the American Chemical Society, 110*(7), 2145–2147. https://doi.org/10.1021/JA00215A023/ASSET/JA00215A023.FP.PNG_V03

82.   Spackman, P. R., Turner, M. J., Mckinnon, J. J., Wolff, S. K., Grimwood, D. J., Jayatilaka, D., & Spackman, M. A. (n.d.). CrystalExplorer: A program for Hirshfeld surface analysis, visualization and quantitative analysis of molecular crystals. *scripts.iucr.org*. https://doi.org/10.1107/S1600576721002910

83.   Seth, S. K., Sarkar, D., & Kar, T. (2011). Use of π-π Forces to steer the assembly of chromone derivatives into hydrogen bonded supramolecular layers: Crystal structures and Hirshfeld surface analyses. *CrystEngComm, 13*(14), 4528–4535. https://doi.org/10.1039/c1ce05037k

84.   Shit, S., Marschner, C., & Mitra, S. (2016). Synthesis, Crystal Structure, and Hirshfeld Surface Analysis of a New Mixed Ligand Copper(II) Complex. *Acta Chimica Slovenica, 63*(1), 129–137. https://doi.org/10.17344/ACSI.2015.2024

85.   Seth, S. K., Sarkar, D., & Kar, T. (2011). Use of π–π forces to steer the assembly of chromone derivatives into hydrogen bonded supramolecular layers: crystal structures and Hirshfeld surface analyses. *CrystEngComm, 13*(14), 4528–4535. https://doi.org/10.1039/C1CE05037K

86.   An, X. X., Zhao, Q., Mu, H. R., & Dong, W. K. (2019). A New Half-Salamo-Based Homo-Trinuclear Nickel(II) Complex: Crystal Structure, Hirshfeld Surface Analysis, and Fluorescence Properties. *Crystals 2019, Vol. 9, Page 101*, *9*(2), 101. https://doi.org/10.3390/CRYST9020101

87.   Praveen, K., Madhavi, D. S. S., Kumar, A. K., & Kranthi Kumar, Y. (2016). Coordination chemistry of salicylic acid. *International Journal of Engineering and Science Invention*, *5*(9), 2319–6734.

88.   Zelenák, V., Györyová, K., & Mlynarcík, D. (2002). Antibacterial and Antifungal Activity of Zinc(II) Carboxylates With/Without N-Donor Organic Ligands. *Metal-based drugs*, *8*(5), 269–274. https://doi.org/10.1155/MBD.2002.269

89.   Cho, J., Moon, J., Bioscience, K. S.-, biotechnology, undefined, & 1998, undefined. (1998). Antimicrobial Activity of 4-Hydroxybenzoic Acid and trans 4-Hydroxycinnamic Acid Isolated and Identified from Rice Hull. *Taylor & Francis*, *62*(11), 2273–2276. https://doi.org/10.1271/bbb.62.2273

90.   Sytar Oksana. (2012). Plant phenolic compounds for food, pharmaceutical and cosmetics production. *Journal of Medicinal Plants Research*, *6*(13), 2526–2538. https://doi.org/10.5897/jmpr11.1695

91.   Chaudhary, J., Jain, A., Manuja, R., & Sachdeva, S. (2013). A comprehensive review on biological activities of p-hydroxy benzoic acid and its derivatives. *researchgate.net*, *22*(2), 109–115. Opgehaal van https://www.researchgate.net/profile/Jasmine-

Chaudhary/publication/264420140_A_Comprehensive_Revie
w_on_Biological_activities_of_p-
hydroxy_benzoic_acid_and_its_derivatives/links/5bbc806792
851c7fde371acb/A-Comprehensive-Review-on-Biological-
activities-of-p-hydroxy-benzoic-acid-and-its-derivatives.pdf

92.   Bergmann, H., & Eckert, H. (1990). Effect of
monoethanolamine on growth and biomass formation of rye
and barley. *Plant Growth Regulation, 9*(1), 1–8.
https://doi.org/10.1007/BF00025273/METRICS

93.   Zardini, H. Z., Davarpanah, M., Shanbedi, M., Amiri, A.,
Maghrebi, M., & Ebrahimi, L. (2014). Microbial toxicity of
ethanolamines - Multiwalled carbon nanotubes. *Journal of
Biomedical Materials Research - Part A, 102*(6), 1774–1781.
https://doi.org/10.1002/JBM.A.34846

94.   Moussa, H. R., El-Sayed Mohamed Selem, E., &
Ghramh, H. A. (2019). Ethanolamine affects physiological
responses of salt-treated jute plants. *International Journal of
Vegetable Science, 25*(6), 581–589.
https://doi.org/10.1080/19315260.2019.1566187

95.   Han, L., Gao, J. R., Li, Z. M., Zhang, Y., & Guo, W. M.
(2007). Synthesis of new plant growth regulator: N-(fatty acid)
O-aryloxyacetyl ethanolamine. *Bioorganic & medicinal
chemistry letters, 17*(11), 3231–3234.
https://doi.org/10.1016/J.BMCL.2007.03.013

96.   Sumalan, R. L., Halip, L., Maffei, M. E., Croitor, L.,
Siminel, A. V., Radulov, I., ... Crisan, M. E. (2021).
Bioprospecting Fluorescent Plant Growth Regulators from

Arabidopsis to Vegetable Crops. *International Journal of Molecular Sciences 2021, Vol. 22, Page 2797, 22*(6), 2797. https://doi.org/10.3390/IJMS22062797

97.   Simina, A., Crisan, M., Ciulca, S., & Botau, D. (2017). The influence of monoethanolamine 4-aminobenzoate on callus growth of Momordica charantia L. *Journal of Horticulture, Forestry and Biotechnology, 21*(1), 174–178.

98.   Ibragimov, A. B., Ashurov, Z. M., & Tashpulatov, Z. Z. (2017). Synthesis, structure, and fungicidal activity of mono- and binuclear mixed-ligand copper complex with p-nitrobenzoic acid and monoethanolamine. *Russian Journal of Coordination Chemistry/Koordinatsionnaya Khimiya, 43*(6), 380–388. https://doi.org/10.1134/S1070328417060021

99.   Seth, S. K., Saha, I., Estarellas, C., Frontera, A., Kar, T., & Mukhopadhyay, S. (2011). Supramolecular self-assembly of M-IDA complexes involving lone-pair ⋯φ Interactions: Crystal structures, hirshfeld surface analysis, and DFT calculations [H2IDA = iminodiacetic acid, M = Cu(II), Ni(II)]. *Crystal Growth and Design, 11*(7), 3250–3265. https://doi.org/10.1021/CG200506Q/SUPPL_FILE/CG200506Q _SI_003.CIF

100.   Al-Anber, M. A., Rüffer, T., Al-Noaimi, M., & Lang, H. (2015). Synthesis, solid-state structure and supramolecularity of [Cu(pyterpy)2](ClO4)2. *Arabian Journal of Chemistry, 8*(5), 678–684. https://doi.org/https://doi.org/10.1016/j.arabjc.2012.02.006

101.   Deng, Q., Xiao, H., Fang, Q., Cai, Q., Han, Y., & Cai, T.

(2019). A solid-state supramolecular hybrid complex based on dimethylammonium ions and silicon molybdate: Crystal structure, photochromic properties and catalytic performance. *Journal of Solid State Chemistry*, *271*, 81–87. https://doi.org/https://doi.org/10.1016/j.jssc.2018.12.046

102.   Zhao, Y., Zou, K.-Q., Zheng, W.-X., Huang, C.-C., Zheng, B.-Y., Ke, M.-R., & Huang, J.-D. (2022). Solid-state supramolecular structures and excellent photothermal activities of dimeric zinc(II) phthalocyanines axially bridged with bipyridine derivatives. *Dyes and Pigments*, *199*, 110037. https://doi.org/https://doi.org/10.1016/j.dyepig.2021.110037

103.   Bhaumik, P. K., Frontera, A., & Chattopadhyay, S. (2021). An insight into the role of supramolecular interactions to stabilize the solid state structure of an octahedral nickel(II) diamine complex. *Inorganica Chimica Acta*, *515*, 120023. https://doi.org/https://doi.org/10.1016/j.ica.2020.120023

104.   Nedelcu, A., Žak, Z., Madalan, A. M., Pinkas, J., & Andruh, M. (2003). Supramolecular solid-state architectures constructed from 4,4'-bipyridine-N,N'-dioxide and dicyanamido tectons. Synthesis and crystal structures of [M(bpno)2{N(CN)2}2(H2O)2] (M=Co, Mn) and [Cu(bpno){N(CN)2}2(H2O)]. *Polyhedron*, *22*(5), 789–794. https://doi.org/https://doi.org/10.1016/S0277-5387(02)01411-0

105.   Rakshit, G., Komal, Dagur, P., Biswas, A., Murtuja, S., & Jayaprakash, V. (2023). Chapter 9 - Molecular docking and molecular dynamics in natural products-based drug discovery.

In C. Egbuna, M. Rudrapal, & H. B. T.-P. Tijjani Computational Tools and Databases in Drug Discovery (Reds), *Drug Discovery Update* (bll 195–212). Elsevier. https://doi.org/https://doi.org/10.1016/B978-0-323-90593-0.00018-6

106.   Sánchez-Cruz, N. (2023). Deep graph learning in molecular docking: Advances and opportunities. *Artificial Intelligence in the Life Sciences*, *3*, 100062. https://doi.org/https://doi.org/10.1016/j.ailsci.2023.100062

107.   Gupta, P. S. Sen, Bhat, H. R., Biswal, S., & Rana, M. K. (2020). Computer-aided discovery of bis-indole derivatives as multi-target drugs against cancer and bacterial infections: DFT, docking, virtual screening, and molecular dynamics studies. *Journal of Molecular Liquids*, *320*, 114375. https://doi.org/https://doi.org/10.1016/j.molliq.2020.114375

108.   Stanzione, F., Giangreco, I., & Cole, J. C. (2021). Chapter Four - Use of molecular docking computational tools in drug discovery. In D. R. Witty & B. B. T.-P. in M. C. Cox (Reds), (Vol 60, bll 273–343). Elsevier. https://doi.org/https://doi.org/10.1016/bs.pmch.2021.01.004

109.   Issa, N. T., Badiavas, E. V, & Schürer, S. (2019). Research Techniques Made Simple: Molecular Docking in Dermatology - A Foray into In Silico Drug Discovery. *Journal of Investigative Dermatology*, *139*(12), 2400-2408.e1. https://doi.org/https://doi.org/10.1016/j.jid.2019.06.129

110.   Okimoto, N., Futatsugi, N., Fuji, H., Suenaga, A., Morimoto, G., Yanai, R., … Taiji, M. (2010). High-Performance

Drug Discovery: Computational Screening by Combining Docking and Molecular Dynamics Simulations. *Biophysical Journal*, *98*(3, Supplement 1), 460a. https://doi.org/https://doi.org/10.1016/j.bpj.2009.12.2502

111.   Asiamah, I., Obiri, S. A., Tamekloe, W., Armah, F. A., & Borquaye, L. S. (2023). Applications of molecular docking in natural products-based drug discovery. *Scientific African*, *20*, e01593. https://doi.org/https://doi.org/10.1016/j.sciaf.2023.e01593

112.   Keshavarz, F., & Barbiellini, B. (2023). Application of molecular docking simulation to screening of metal–organic frameworks. *Computational Materials Science*, *226*, 112257. https://doi.org/https://doi.org/10.1016/j.commatsci.2023.11 2257

113.   Abdolmaleki, A., Shiri, F., & Ghasemi, J. B. (2021). Chapter 11 - Use of Molecular Docking as a Decision-Making Tool in Drug Discovery. In M. S. B. T.-M. D. for C.-A. D. D. Coumar (Red), (bll 229–243). Academic Press. https://doi.org/https://doi.org/10.1016/B978-0-12-822312-3.00010-2

114.   Ajala, A., Uzairu, A., Shallangwa, G. A., Abechi, S. E., Ramu, R., & Al-Ghorbani, M. (2023). Natural product inhibitors as potential drug candidates against Alzheimer's disease: Structural-based drug design, molecular docking, molecular dynamic simulation experiments, and ADMET predictions. *Journal of the Indian Chemical Society*, *100*(5), 100977.

https://doi.org/https://doi.org/10.1016/j.jics.2023.100977

115.   Wingbermühle, S., & Lindahl, E. (2023). Fully automated screening of compound libraries in drug discovery using docking and molecular dynamics. *Biophysical Journal*, *122*(3, Supplement 1), 182a. https://doi.org/https://doi.org/10.1016/j.bpj.2022.11.1126

116.   Adelusi, T. I., Oyedele, A.-Q. K., Boyenle, I. D., Ogunlana, A. T., Adeyemi, R. O., Ukachi, C. D., … Abdul-Hammed, M. (2022). Molecular modeling in drug discovery. *Informatics in Medicine Unlocked*, *29*, 100880. https://doi.org/https://doi.org/10.1016/j.imu.2022.100880

117.   Ganugapati, J., & Akash, S. (2017). Multi-template homology based structure prediction and molecular docking studies of protein 'L' of Zaire ebolavirus (EBOV). *Informatics in Medicine Unlocked*, *9*, 68–75. https://doi.org/https://doi.org/10.1016/j.imu.2017.06.002

118.   Takaoka, Y., Sugano, A., Morinaga, Y., Ohta, M., Miura, K., Kataguchi, H., … Maniwa, Y. (2022). Prediction of infectivity of SARS-CoV2: Mathematical model with analysis of docking simulation for spike proteins and angiotensin-converting enzyme 2. *Microbial Risk Analysis*, *22*, 100227. https://doi.org/https://doi.org/10.1016/j.mran.2022.100227

119.   Ali Eltayb, W., Abdalla, M., Ahmed EL-Arabey, A., Boufissiou, A., Azam, M., Al-Resayes, S. I., & Alam, M. (2023). Exploring particulate methane monooxygenase (pMMO) proteins using experimentation and computational molecular docking. *Journal of King Saud University - Science*, *35*(4),

102634.
https://doi.org/https://doi.org/10.1016/j.jksus.2023.102634

120.   Fanelli, A., & Sullivan, M. L. (2023). Chapter Three - Tools for protein structure prediction and for molecular docking applied to enzyme active site analysis: A case study using a BAHD hydroxycinnamoyltransferase. In J. B. T.-M. in E. Jez (Red), *Biochemical Pathways and Environmental Responses in Plants: Part C* (Vol 683, bll 41–79). Academic Press. https://doi.org/https://doi.org/10.1016/bs.mie.2022.10.004

121.   Arputharaj, D. S., Rajasekaran, M., & Nidhin, P. V. (2023). Sulfamethoxazole: Molecular docking and crystal structure prediction. *Results in Chemistry, 5*, 100716. https://doi.org/https://doi.org/10.1016/j.rechem.2022.100716

122.   AL-Refaei, M. A., Makki, R. M., & Ali, H. M. (2020). Structure prediction of transferrin receptor protein 1 (TfR1) by homology modelling, docking, and molecular dynamics simulation studies. *Heliyon, 6*(1), e03221. https://doi.org/https://doi.org/10.1016/j.heliyon.2020.e03221

123.   Shao, L., Wang, J., Hu, H., Xu, X., & Wang, H. (2023). The interaction of an effector protein Hap secreted by Aeromonas salmonicida and myofibrillar protein of meat: Possible mechanisms from structural changes to sites of molecular docking. *Food Chemistry, 424*, 136365. https://doi.org/https://doi.org/10.1016/j.foodchem.2023.136

124.	Hasan, S. N., Banerjee, J., Patra, S., Kar, S., Das, S., Samanta, S., … Bag, B. G. (2023). Self-assembled renewable nano-sized pentacyclic triterpenoid maslinic acids in aqueous medium for anti-leukemic, antibacterial and biocompatibility studies: An insight into targeted proteins-compound interactions based mechanistic pathway prediction through molecular docking. *International Journal of Biological Macromolecules, 245*, 125416. https://doi.org/https://doi.org/10.1016/j.ijbiomac.2023.125416

125.	Mijit, A., Wang, X., Li, Y., Xu, H., Chen, Y., & Xue, W. (2023). Mapping synthetic binding proteins epitopes on diverse protein targets by protein structure prediction and protein-protein docking. *Computers in Biology and Medicine, 163*, 107183. https://doi.org/https://doi.org/10.1016/j.compbiomed.2023.107183

126.	Costa, L. D., Silva, C. F. M., Pinto, D. C. G. A., Silva, A. M. S., Pereira, F., Faustino, M. A. F., & Tomé, A. C. (2023). Discovery of thiazolo[5,4-c]isoquinoline based compounds as acetylcholinesterase inhibitors through computational target prediction, molecular docking and bioassay. *Journal of Molecular Structure*, 136088. https://doi.org/https://doi.org/10.1016/j.molstruc.2023.136088

127.	Crampon, K., Giorkallos, A., Deldossi, M., Baud, S., &

Steffenel, L. A. (2022). Machine-learning methods for ligand–protein molecular docking. *Drug Discovery Today*, *27*(1), 151–164. https://doi.org/https://doi.org/10.1016/j.drudis.2021.09.007

128.   Menezes, T. M., de Almeida, S. M. V., de Moura, R. O., Seabra, G., de Lima, M. do C. A., & Neves, J. L. (2019). Spiro-acridine inhibiting tyrosinase enzyme: Kinetic, protein-ligand interaction and molecular docking studies. *International Journal of Biological Macromolecules*, *122*, 289–297. https://doi.org/https://doi.org/10.1016/j.ijbiomac.2018.10.175

129.   Tumskiy, R. S., Tumskaia, A. V, Klochkova, I. N., & Richardson, R. J. (2023). SARS-CoV-2 proteases Mpro and PLpro: Design of inhibitors with predicted high potency and low mammalian toxicity using artificial neural networks, ligand-protein docking, molecular dynamics simulations, and ADMET calculations. *Computers in Biology and Medicine*, *153*, 106449. https://doi.org/https://doi.org/10.1016/j.compbiomed.2022.106449

130.   Firus Khan, A. Y., Abdul Wahab, R., Ahmed, Q. U., Khatib, A., Ibrahim, Z., Nipun, T. S., … AlKahtane, A. A. (2023). Potential anticancer agents identification of Hystrix brachyura bezoar through gas chromatography-mass spectrometry-based metabolomics and protein-ligand interaction with molecular docking analyses. *Journal of King Saud University - Science*, *35*(6), 102727. https://doi.org/https://doi.org/10.1016/j.jksus.2023.102727

131.   Srinivasa, C., Shivamallu, C., Kallimani, S., Sushma, P., Kollur, S. P., Prabhurajeshwar, & Gopinath, S. M. (2022). Chapter 11 - Application of molecular docking and dynamics tools in SARS-CoV-2 drug design: ligand–protein interaction studies. In C. B. T.-C. D. D. Egbuna (Red), *Drug Discovery Update* (bll 253–271). Elsevier. https://doi.org/https://doi.org/10.1016/B978-0-323-95578-2.00003-0

132.   Kyhoiesh, H. A. K., & Al-Adilee, K. J. (2023). Pt(IV) and Au(III) complexes with tridentate-benzothiazole based ligand: synthesis, characterization, biological applications (antibacterial, antifungal, antioxidant, anticancer and molecular docking) and DFT calculation. *Inorganica Chimica Acta*, *555*, 121598. https://doi.org/https://doi.org/10.1016/j.ica.2023.121598

133.   Mohammadlou, F., Mansouri-Torshizi, H., Dehghanian, E., Eslami-Moghadam, M., Dusek, M., & Eigner, V. (2023). A new zinc(II) complex of 2-benzoimidazoledisulfide ligand: synthesis, X-ray crystallographic structure, investigation of CT-DNA and BSA interaction by spectroscopic techniques and molecular docking. *Journal of Photochemistry and Photobiology A: Chemistry*, *443*, 114830. https://doi.org/https://doi.org/10.1016/j.jphotochem.2023.1 14830

134.   Sahu, S. N., Satpathy, S. S., Pattnaik, S., Mohanty, C., & Pattanayak, S. K. (2022). Boerhavia diffusa plant extract can be a new potent therapeutics against mutant nephrin protein responsible for type1 nephrotic syndrome: Insight into

hydrate-ligand docking interactions and molecular dynamics simulation study. *Journal of the Indian Chemical Society*, *99*(10), 100669. https://doi.org/https://doi.org/10.1016/j.jics.2022.100669

135.   Phusi, N., Sato, R., Ezawa, T., Tomioka, S., Hanwarinroj, C., Khamsri, B., … Kurita, N. (2019). Specific interactions between 2-trans enoyl-acyl carrier protein reductase and its ligand: Protein-ligand docking and ab initio fragment molecular orbital calculations. *Journal of Molecular Graphics and Modelling*, *88*, 299–308. https://doi.org/https://doi.org/10.1016/j.jmgm.2019.02.011

136.   Wu, Z., Xu, J., Ruan, J., Chen, J., Li, X., Yu, Y., … Li, H. (2023). Probing the mechanism of interaction between capsaicin and myofibrillar proteins through multispectral, molecular docking, and molecular dynamics simulation methods. *Food Chemistry: X*, *18*, 100734. https://doi.org/https://doi.org/10.1016/j.fochx.2023.100734

137.   Konovalov, B., Đorđević, I. S., Franich, A. A., Šmit, B., Živković, M. D., Djuran, M. I., … Rajković, S. (2023). Dinuclear platinum(II) complexes with 1,5-nphe bridging ligand: Spectroscopic and molecular docking study of the interactions with N-acetylated L-methionylglycine and human serum albumin. *Journal of Molecular Structure*, *1288*, 135810. https://doi.org/https://doi.org/10.1016/j.molstruc.2023.135810

138.   Li, Y., Li, Y., Liu, X., Yang, Y., Lin, D., & Gao, Q. (2020). The synthesis, characterization, DNA/protein interaction,

molecular docking and catecholase activity of two Co(II) complexes constructed from the aroylhydrazone ligand. *Journal of Molecular Structure, 1202*, 127229. https://doi.org/https://doi.org/10.1016/j.molstruc.2019.127229

139.	Murugesu, S., Ibrahim, Z., Ahmed, Q. U., Uzir, B. F., Nik Yusoff, N. I., Perumal, V., … Khatib, A. (2019). Identification of α-glucosidase inhibitors from Clinacanthus nutans leaf extract using liquid chromatography-mass spectrometry-based metabolomics and protein-ligand interaction with molecular docking. *Journal of Pharmaceutical Analysis, 9*(2), 91–99. https://doi.org/https://doi.org/10.1016/j.jpha.2018.11.001

140.	Dubey, S. M., Han, S., Stutzman, N., Prigge, M. J., Medvecká, E., Platre, M. P., … Estelle, M. (2023). The AFB1 auxin receptor controls the cytoplasmic auxin response pathway in Arabidopsis thaliana. *Molecular Plant*. https://doi.org/https://doi.org/10.1016/j.molp.2023.06.008

141.	Jiang, Y., Wang, Z., Wu, Y., Li, H., & Xue, X. (2023). The effect of auxin status driven by bacterivorous nematodes on root growth of Arabidopsis thaliana. *Applied Soil Ecology, 182*, 104730. https://doi.org/https://doi.org/10.1016/j.apsoil.2022.104730

142.	Djemal, R., Bradai, M., Amor, F., Hanin, M., & Ebel, C. (2023). Wheat type one protein phosphatase promotes salt and osmotic stress tolerance in arabidopsis via auxin-mediated remodelling of the root system. *Plant Physiology and Biochemistry, 201*, 107832.

https://doi.org/https://doi.org/10.1016/j.plaphy.2023.107832

143.   Araniti, F., Talarico, E., Madeo, M. L., Greco, E., Minervino, M., Álvarez-Rodríguez, S., … Bruno, L. (2023). Short-term exposition to acute cadmium toxicity induces the loss of root gravitropic stimuli perception through PIN2-mediated auxin redistribution in Arabidopsis thaliana (L.) Heynh. *Plant Science, 332*, 111726. https://doi.org/https://doi.org/10.1016/j.plantsci.2023.111726

144.   Ma, M., Lu, Y., Di, D., Kronzucker, H. J., Dong, G., & Shi, W. (2023). The nitrification inhibitor 1,9-decanediol from rice roots promotes root growth in Arabidopsis through involvement of ABA and PIN2-mediated auxin signaling. *Journal of Plant Physiology, 280*, 153891. https://doi.org/https://doi.org/10.1016/j.jplph.2022.153891

www.ingramcontent.com/pod-product-compliance
Lightning Source LLC
Chambersburg PA
CBHW070842160726

48004CB00001B/469